NUMBER POWER

A REAL WORLD APPROACH TO MATH

Graphs, Charts, Schedules, and Maps

ROBERT MITCHELL AND DONALD PRICKEL

McGraw Hill **Education**

Bothell, WA • Chicago, IL • Columbus, OH • New York, NY

www.mheonline.com

Send all inquiries to:
Contemporary/McGraw-Hill
130 E. Randolph, Suite 400
Chicago, IL 60601

ISBN: 978-0-07-659230-2
MHID: 0-07-659230-8

Printed in the United States of America.

4 5 6 7 8 9 RHR 15 14 13

The McGraw·Hill Companies

TABLE OF CONTENTS

Schedules and Charts

Maps

Posttest A

Using Number Power   **153**

TO THE STUDENT

Welcome to *Graphs, Charts, Schedules, and Maps:*

This book introduces you to the skills you need to read a wide variety of graphs, charts, schedules, and maps. An understanding of these topics is necessary in many occupations and in your everyday life. Also, reading graphs, charts, schedules, and maps is a standard section on educational and vocational tests, including GED, college entrance, civil service, and military tests.

The first section of this book, Building Number Power, provides step-by-step instruction in reading and interpreting these kinds of materials. You will learn to find information from four kinds of graphs: pictographs, circle graphs, line graphs, and bar graphs. You will practice reading a variety of schedules and charts that you might face on a daily basis. And you will learn how to interpret information from three types of maps: geographical maps, directional maps, and informational maps.

Each chapter begins with a skills inventory to help identify what you need to learn. After completing the lessons and exercises, you will be challenged in applying those skills to everyday problems. Each chapter ends with a skills review to check your progress on what you have just learned.

Learning how and when to use a calculator is an important skill to develop. In real life, problem solving often involves the smart use of a calculator – especially when working with large numbers. You need to know when an answer on a calculator makes sense, so be sure to have an estimated answer in mind. Remember that a calculator can help you only if you set up the problem and use the calculator correctly.

To get the most out of your work, do each problem carefully. Check each answer to make sure you are working accurately. An answer key is provided at the back of the book. Inside the back cover is a chart to help you keep track of your score in each exercise.

Pretest

This test will tell you which sections of this book you need to concentrate on. Do every problem that you can. Correct answers are listed by page number at the back of the book. After you check your answers, the chart at the end of the test will guide you to the pages of the book where you need work.

GRAPH A

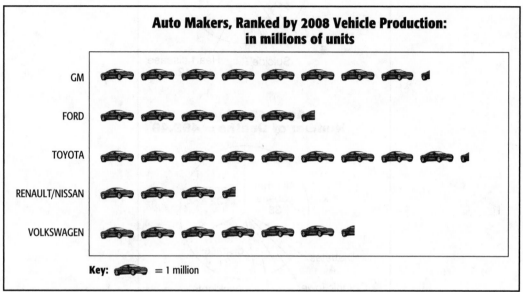

1. Graph A shows the world vehicle production by the _____ of cars with the symbol representing _____ cars.

2. In 2008, Toyota produced almost $5\frac{1}{2}$ million cars.   True   False

3. According to Graph A, the total production of Ford is slightly greater than _____ the combined production of Nissan and Volkswagen.
   a. $\frac{1}{2}$ b. $\frac{4}{5}$ c. $\frac{1}{3}$ d. $\frac{2}{3}$ e. 2x (twice)

4. Which statement best describes Graph A?
   a. Toyota is catching up with Ford in car production.
   b. Volkswagen has had a bad year.
   c. In 2008, Toyota was the top producing manufacturer.
   d. Americans like to drive American cars.
   e. 31,300,000 cars are produced every year.

GRAPH B

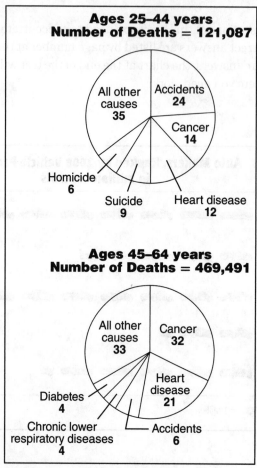

Ages 25–44 years
Number of Deaths = 121,087

- All other causes 35
- Accidents 24
- Cancer 14
- Heart disease 12
- Suicide 9
- Homicide 6

Ages 45–64 years
Number of Deaths = 469,491

- All other causes 33
- Cancer 32
- Heart disease 21
- Accidents 6
- Chronic lower respiratory diseases 4
- Diabetes 4

Source: National Vital Statistics System, Mortality

5. Graph B shows the percent distribution of the _____ leading causes of _____.

6. Cancer is the leading cause of death for both age groups.　　　　　　　True　　　False

7. _____ of all deaths for persons aged 25–44 years are due to a combination of accidents, cancer, and heart disease.

　　a. 33%　　　　　　　　d. 20%
　　b. 50%　　　　　　　　e. 15%
　　c. 75%

8. It can be concluded from Graph B that all of the following are true EXCEPT:

　　a. The three leading causes of death in both age groups are the same.
　　b. The five leading causes of death are not the same for each age group.
　　c. Twenty-nine thousand and sixty persons die annually from accidents in the age 25–44 year old category.
　　d. Guns are the causes of suicide and homicide deaths in 25–44 year olds.
　　e. There are almost 4 times the number of deaths in 45–64 years olds as there are in 25–44 years olds.

GRAPH C

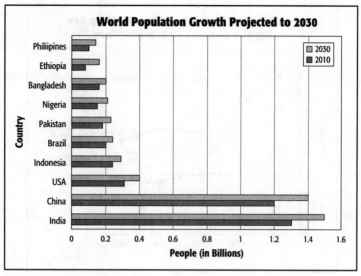

Source: U.S. Census Bureau

9. Graph C shows _____ in the world projected to _____.

10. The greatest population growth is projected to occur in the United States.　　　True　　　False

11. What country will double its population between 2010 and 2030?

　　a. India
　　b. United States
　　c. Ethiopia
　　d. China
　　e. Brazil

12. Which statement best summarizes the information on Graph C?

　　a. United States and Brazil will soon have the same population.
　　b. The world's population will continue to increase and grow
　　　　through 2030.
　　c. China's population is growing at a faster rate than that
　　　　of any other country.
　　d. The population of the world between 2010 and 2030 will
　　　　probably double.

GRAPH D

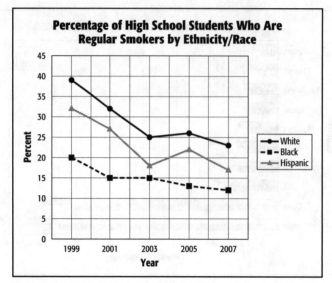

Source: Centers for Disease Control and Prevention, 2008

13. Graph D shows the percentage of regular smokers who are _____ from 1999 to _____.

14. In 1999, white students were smoking as much as Hispanic students.　　　　　True　　　　False

15. In 2007 what was the difference in percentage between white and black high school students who were regular smokers?

　　a. less than 5%

　　b. more than 10%

　　c. 4.0%

　　d. more than 20%

　　e. less than 7%

16. From the information on Graph D you could infer that

　　a. the campaign against smoking is working

　　b. teenagers are smoking at an earlier age

　　c. black students smoke more frequently than Hispanic students

　　d. smoking is increasing rapidly among Hispanic students

　　e. smoking among all high school students seems to be leveling off
　　　if not decreasing

CHART A

AVERAGE UNEMPLOYMENT FOR 2010		
	Ages	
	16–19	20 and older
Black men	22.3%	19%
Black women	18.8%	12.4%
White men	12.2%	8.9%
White women	11.5%	7.3%

Source: Bureau of Labor Statistics, 2010

17. Chart A shows percentages of _____ in 2010 by race, sex, and age.

18. By looking at Chart A, you can compare the difference in joblessness for people from teenage years to retirement.   True   False

19. The average unemployment rate for black men aged 20 and older is _____ than that of white men of the same age.

    a. more than two times less
    b. more than two times greater
    c. three times less
    d. over three times greater
    e. more than one-half greater

20. The following statement is true according to Chart A:

    a. More elderly people are unemployed than teenagers.
    b. Women have less trouble getting jobs than men.
    c. Black men consistently have higher unemployment rates than white men of the same age.
    d. The reason that white teenage women have low unemployment rates is because they stay in school.
    e. Unemployment is cut by half the older you get.

SCHEDULE A

City of New Orleans

59		◀ Train Number ▶		58
Daily		◀ Days of Operation ▶		Daily
Read Down	▼		▲	Read Up
8:00P	Dp	Chicago, IL–Union Station	Ar	9:20A
8:51P		Homewood, IL		8:14A
9:24P		Kankakee, IL		7:39A
10:37P		Champaign-Urbana, IL		6:33A
11:18P		Mattoon, IL		5:44A
11:43P		Effingham, IL		5:17A
12:35A		Centralia, IL		4:28A
1:30A 1:35A	Ar Dp	Carbondale, IL	Dp Ar	3:33A 3:28A
3:40A		Fulton, KY		1:26A
4:25A		Newbern-Dyersburg, TN		12:36A
6:29A 6:45A	Ar Dp	Memphis, TN	Dp Ar	10:53P 10:41P
9:00A		Greenwood, MS		8:06P
9:59A		Yazoo City, MS		7:06P
11:09A		Jackson, MS		6:06P
11:48A		Hazlehurst, MS		5:14P
12:10P		Brookhaven, MS		4:53P
12:39P		McComb, MS		4:26P
1:45P		Hammond, LA		3:24P
3:30P	Ar	New Orleans, LA	Dp	2:15P

21. Schedule A shows the train schedule for Trains 59 and 58 which travel
between _____ and _____.

22. The train called *City of New Orleans* travels through the states of Illinois,   True   False
Kentucky, Tennessee, Mississippi, and Louisiana.

23. At what time does Train 59 going to New Orleans leave Greenwood, Mississippi?

    a. 9:00 P.M.   **d.** 9:00 A.M.

    b. 8:00 P.M.   **e.** none of the above

    c. 8:00 A.M.

24. If you left New Orleans at 2:15 in the afternoon, you would arrive in Chicago at

    a. 9:00 the next evening   **d.** 9:20 the next morning

    b. 9:20 the next evening   **e.** 8:00 at night

    c. 9:20 the same morning

MAP A

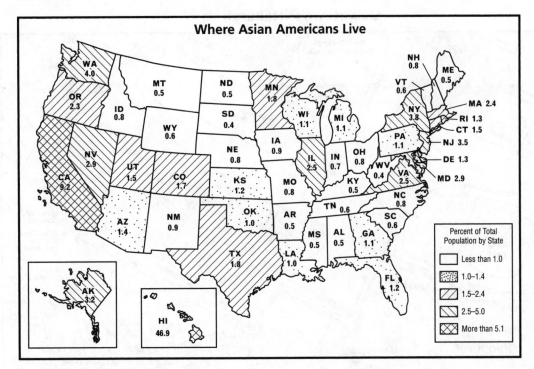

Where Asian Americans Live

Percent of Total Population by State

□	Less than 1.0
▦	1.0–1.4
▨	1.5–2.4
▧	2.5–5.0
▩	More than 5.1

25. The map pictured above shows where _____ live in the United States.

26. According to the map, the states with the largest percentage of the Asian American population are California and Hawaii.   True   False

27. The percentage of Asian Americans in each state with the checkerboard design is _____ more than states not shaded at all.

    a. 3 times

    b. $1\frac{1}{2}$ times

    c. $2\frac{1}{2}$ times

    d. 5 times

    e. 4 times

28. Which statement best summarizes the information on the map?

    a. The New England states each have about the same number of Asian Americans living there.
    b. Most Asian Americans living in the U.S. are Chinese.
    c. The largest concentration of Asian Americans in the U.S. are in states bordering the Atlantic Ocean.
    d. California has a larger percentage of Asian Americans in its population than New Mexico does.
    e. Asian Americans live in states where the climate is warm.

MAP B

Downtown Pittsburgh

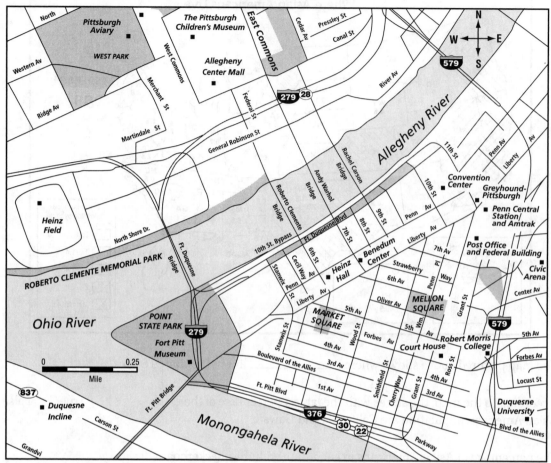

29. The rivers that join together in downtown Pittsburgh are the _____, _____, and _____.

30. It is more than a 5-mile drive between the Aviary in West Park and Fort Pitt Museum in Point State Park.   True   False

31. If you drive north on U.S. Interstate 579 from Duquesne University and go west on U.S. Interstate 279, and then turn north on Federal Street, you will come to what point of interest?

    a. Civic Arena
    b. Heinz Field
    c. Pittsburgh Aviary
    d. Allegheny Center Mall

32. The directional map of downtown Pittsburgh contains the following information:

    a. location of airports
    b. location of hotels
    c. location of points of interest
    d. location of public libraries

MAP C

United States

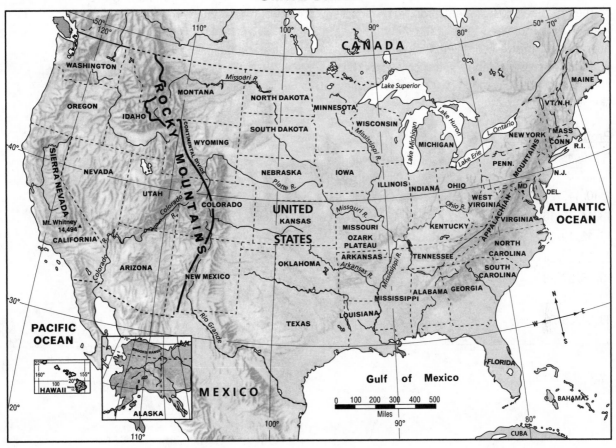

33. Map C shows the United States, which is bordered by the countries _____ and _____.

34. The major mountain range in the eastern section of the United States is called the Sierra Nevada.   True   False

35. What are the two larger mountain ranges located west of the Mississippi River?

    a. Alaskan and Sierra Nevada   **d.** Sierra Nevada and Appalachian
    b. Sierra Nevada and Ozark Plateau   **e.** Rocky and Appalachian
    c. Rocky and Sierra Nevada

36. It can be concluded from Map C that all of the following are true EXCEPT:

    a. Austin is the state capital of Texas.
    b. Five great lakes lie between the northeast border of the
       United States and Canada.
    c. The eastern section of the United States possesses mountainous areas.
    d. The continental divide runs through the Rocky Mountains.
    e. The United States has many rivers and mountains.

PRETEST CHART

If you miss more than one problem in any section of this test, you should complete the lessons on the practice pages indicated on this chart. If you miss only one problem in a section of this test, you may not need further study in that chapter. However, before you skip those lessons, we recommend that you complete the review test at the end of that chapter. For example, if you miss one problem about graphs, you should pass the Graph Review (pages 68–76) before beginning the chapter on schedules and charts. This longer inventory will be a more precise indicator of your skill level.

Problem Numbers	Skill Area	Practice Pages
1, 2, 3, 4	pictograph	20–31
5, 6, 7, 8	circle graph	32–43
9, 10, 11, 12	bar graph	44–55
13, 14, 15, 16	line graph	56–67
17, 18, 19, 20	chart	80–87, 92–95
21, 22, 23, 24	schedule	80–85, 88–95
25, 26, 27, 28	informational map	130–134
29, 30, 31, 32	directional map	124–129
33, 34, 35, 36	geographical map	118–123

BUILDING
NUMBER
POWER

GRAPHS

Graph Skills Inventory

The Graph Skills Inventory allows you to measure your skills in reading and interpreting graphs. Correct answers are listed at the back of the book.

GRAPH A

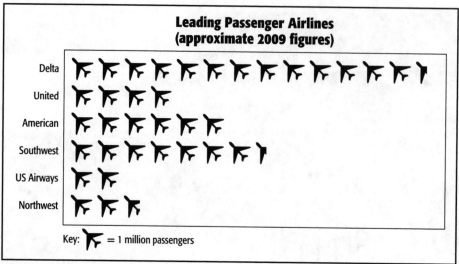

Source: Air Transport Association of America, Inc.

Answer each question below.

1. The two airlines that carry the greatest number of passengers are Delta and _____.

2. Delta Airlines carries approximately _____ passengers each year.

3. Graph A represents the number of passengers carried on the six leading passenger airlines in 2009.   True   False

4. United Airlines and American Airlines carried approximately the same number of passengers.   True   False

5. What was the approximate total number of passengers carried by the six leading airlines in 2009?

   a. 30.5 million   **c.** 35.75 million   **e.** 13.5 million
   b. 10.75 million   **d.** 45 million

6. Delta carried more than _____ as many passengers as American.

   a. one-half   **c.** one-fourth   **e.** three times
   b. twice   **d.** four times

GRAPH B

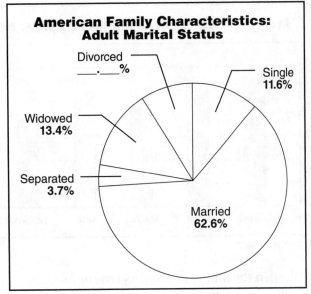

**American Family Characteristics:
Adult Marital Status**

Divorced
___.___%

Single
11.6%

Widowed
13.4%

Separated
3.7%

Married
62.6%

Source: U.S. Census Bureau, 2008

Answer each question by filling in the blank, answering true or false, or by choosing the best multiple-choice response.

7. The categories of marital status shown are married, separated, _____, divorced, and _____.

8. The percent of the American population who are divorced is _____. (Fill this in on the graph.)

9. More adults are separated and divorced than are single.   True   False

10. More than half of adult Americans are married.   True   False

11. What is the percent of adult Americans who are not single?
    a. 74.1%
    b. 11.6%
    c. 88.4%
    d. 50%
    e. 8.7%

12. About one-fourth of all adult Americans are
    a. single or separated
    b. single or widowed
    c. married
    d. divorced or single
    e. widowed or separated

GRAPH C

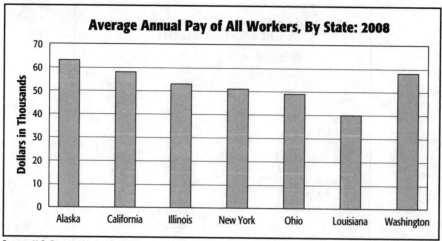

Average Annual Pay of All Workers, By State: 2008

Source: U.S. Bureau of Labor Statistics

Answer each question by filling in the blank, answering true or false, or choosing the best multiple-choice response.

13. Graph C shows the median annual pay for all workers in different _____.

14. The median annual pay in Louisiana was _____.

15. The median annual pay for all workers is shown for the states of New York and Florida.   True   False

16. The median annual pay in California was $32,000.   True   False

17. The median annual pay in Ohio is _____ than the average annual pay in Alaska.

   a. $14,000 more
   b. $8,000 less
   c. $6,000 less
   d. $11,000 less
   e. $14,000 less

18. Approximately what is the median annual pay for all workers in the seven states combined?

   a. $58,400
   b. $53,100
   c. $45,400
   d. $61,900
   e. $50,000

GRAPH D

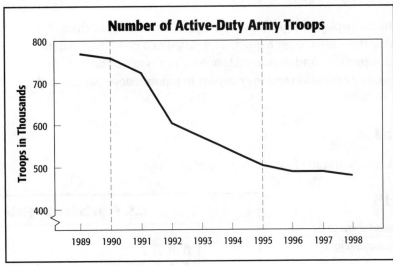

Source: U.S. Department of Defense

Answer each question below.

19. In 1992 approximately _____ thousand troops were active in the U.S. Army.

20. From 1992 to 1998, there was a decrease of about 100,000 troops.　　　　True　　　　False

21. Between which years was the smallest decrease of Army troops?

　　a. 1990–1994
　　b. 1995–1998
　　c. 1992–1995

GRAPH SKILLS INVENTORY CHART

Use this inventory to see what you already know about graphs and what you need to work on. A passing score is 18 correct answers. Even if you have a passing score, circle the number of any problem that you miss and turn to the practice pages indicated for further instruction.

Problem Numbers	Skill Area	Practice Pages
1, 2, 3, 4, 5, 6	pictograph	16–31
7, 8, 9, 10, 11, 12	circle graph	16–19, 32–43
13, 14, 15, 16, 17, 18	bar graph	16–19, 44–55
19, 20, 21	line graph	16–19, 56–67

What Are Graphs?

A **graph** is a pictorial display of information. Since it is drawn, rather than written, a graph makes it possible to get a quick impression of a great deal of data and to easily make comparisons and draw conclusions. Graphs are often used in government, business, and education and may appear in reports, newspapers, and magazines.

Types of Graphs

In this workbook, you will study four main types of graphs.

PICTOGRAPHS

A **pictograph** uses pictures or symbols to display information.

A pictograph usually has a **key** to show the value of each symbol.

Pictographs are read by counting the **symbols** on a line of a graph and computing their value.

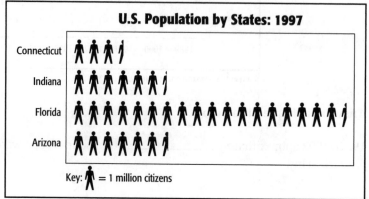

Source: U.S. Census Bureau

CIRCLE GRAPHS

A **circle graph** uses parts of a circle to show information.

Circle graphs show values in each part of a divided circle. A part of a circle graph is called a **segment** or a **section.**

The segments of a circle add up to a whole or to 100% of the topic.

THE GOVERNMENT DOLLAR

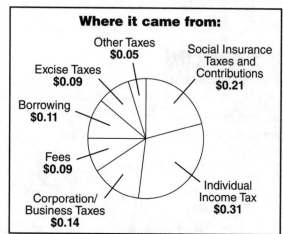

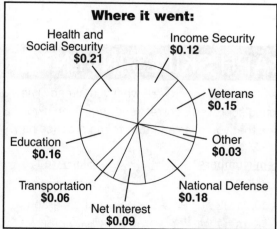

Source: U.S. Office of Management and Budget (2010)

BAR GRAPHS

A **bar graph** uses thick bars to show information.

Bar graphs are usually drawn in one of two different directions:

1. With the bars running up and down. The bars are placed at equal distances along the **horizontal axis** that runs across the bottom of the graph.
2. With the bars running from left to right. The bars are placed at equal distances along the **vertical axis** on the left side of the graph.

To measure a bar accurately, you may want to lay the edge of a sheet of paper across the top of the bar to align with its value at the left (or at the bottom) of the graph.

Sometimes, a graph may have a break in the vertical axis and an open space running across the graph. This means that some values have been left off to save space on the graph. The top graph shows an example of this.

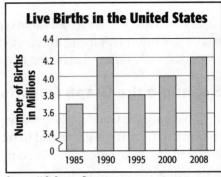

Live Births in the United States

Source: U.S. Census Bureau

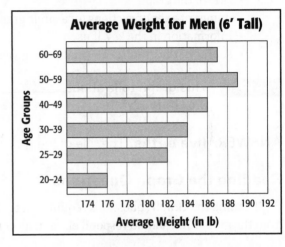

Average Weight for Men (6' Tall)

LINE GRAPHS

A **line graph** is drawn with one or more thin lines that extend across the graph.

Like the bar graph, a line graph is drawn using values along a horizontal and a vertical axis. Using the edge of a paper will help you measure locations on the graph accurately.

A line graph is most useful in showing trends and developments.

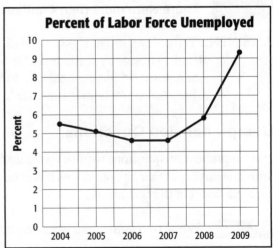

Percent of Labor Force Unemployed

Source: Bureau of Labor Statistics

Types of Questions

In this book, you are asked three types of questions. These questions will help you to find information and interpret graphs. Similar questions will be used in the sections on schedules, charts, and maps.

Scanning the Graph Questions

"Scanning the Graph" questions require you to look carefully at the graph and to fill in missing words to complete a sentence. To complete these statements, pay attention to the

- title of the graph
- names of axes or sections (of a circle graph)
- information in the key, if used
- source of the graph's information, if given

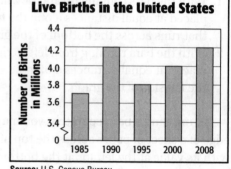

Live Births in the United States

Source: U.S. Census Bureau

EXAMPLE   This graph tells about _____ in the United States.

ANSWER: **live births.** The title of the graph gives this information.

Reading the Graph Questions

"Reading the Graph" questions require you to determine specific information from the graph. You should respond either true or false to the given statements.

EXAMPLE   There were more than 4 million live births in 1990.   True   False

ANSWER: **True.** Starting at the bottom of the graph (horizontal axis), find 1990. Then look at the top of the 1990 bar and follow a line across to a point on the vertical axis to the left. You can see that the 1990 bar goes higher than 4.00.

Comprehension Questions

Comprehension questions require you to do computations, make inferences, draw conclusions, or make predictions based on the graph's information. You will circle the letter of the best choice.

EXAMPLE   What two years had approximately the same number of births?

   a. 1990 and 1995
   b. 1990 and 2008
   c. 1985 and 1990
   d. 1995 and 2000
   e. 2000 and 2008

ANSWER: **b. 1990 and 2008.** Compare the heights of the bars. These bars are closer to the same height than any other two bars.

Read Carefully to Avoid Mistakes

Since interpretation is very important when answering questions, be sure to read a question carefully before choosing your response. Some questions may be tricky. The categories below are possible problems to avoid.

Information Is True but Not Contained on the Graph

EXAMPLE True or False? According to Graph A, unemployment rates reflect the conditions of the U.S. economy.

Answer: **False.** While this fact may be true, the graph gives you no information about our country's economy.

Misleading Words Are Given in Answer Choices

EXAMPLE Which statement is true according to Graph A?

   a. The percentage of unemployed workers was greatest in 2008.

   b. The percentage of unemployed workers rose during 2007 to 2009.

   c. The percentage of the labor force unemployed has ranged from 5.5 to 10 percent.

ANSWER: **b.** The graph shows a line representing an increase in percentages of the labor force unemployed from 2007 and continuing to 2009.

   Answers **a.** and **c.** are not true; 2009 shows the largest unemployment percentage, and the range of unemployment has ranged from 5.1 to 9.3, not 10.0 percent.

GRAPH A

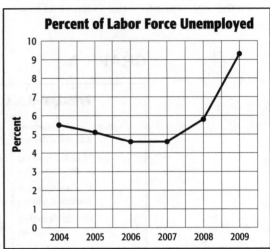

Percent of Labor Force Unemployed

Source: Bureau of Labor Statistics

Labels Are Important

EXAMPLE In 2005, what was the percentage of the labor force unemployed?

   a. 4.6 %
   b. 9.3 %
   c. 5.8%
   d. 5.5%
   e. 5.1%

ANSWER: **e. 5.1 %.** Find 2005 on the horizontal axis. Follow up to the line and look across to the vertical axis to the left. You'll see a value of slightly above 5.0, or approximately 5.1%.

Pictographs

A **pictograph** gets its name from the small pictures it uses as symbols on the graph. Pictographs generally use a **key** to show the value of the pictures that are used as symbols. Parts of symbols are often used to represent a fractional amount of a quantity shown in the key.

Pictographs are often not as exact as other types of graphs, but they are the easiest to read.

GRAPH A

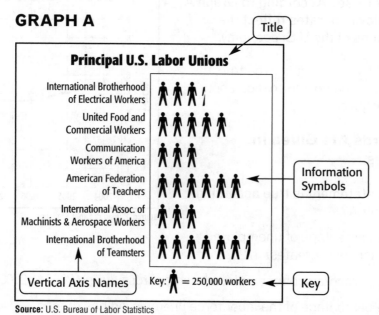

Source: U.S. Bureau of Labor Statistics

To answer questions about a pictograph, follow the sequence below.

Scanning the Graph

To scan a pictograph, find the graph title, the vertical axis names, and the key.

EXAMPLE   The Communication Workers of America is one of the principal _____.

STEP 1   Find the name "Communication Workers of America" on the vertical axis.

STEP 2   Look at the graph title: "Principal U.S. Labor Unions." This title gives the subject of the graph.

ANSWER: **U.S. labor unions**

Reading the Graph

To read a pictograph, count the number of symbols on a line. Then multiply the number of symbols by the value of the symbol given in the key.

Sometimes only a part of the symbol is shown. Look at the partial symbol carefully. Most often, a partial symbol will be $\frac{1}{2}$ of the whole or sometimes $\frac{1}{4}$ or $\frac{3}{4}$. To find a value for a part of a symbol, find that fraction of the whole.

 EXAMPLE   The total membership of workers in the International   True   False
Brotherhood of Teamsters is 1,625,000.

       STEP 1   Find the name "International Brotherhood of Teamsters."

       STEP 2   Count the number of complete symbols: six (6) Next, determine the fraction that the partial symbol represents: one-half symbol $\left(\frac{1}{2}\right)$. There are $6\frac{1}{2}$ symbols.

       STEP 3   Compute the value of the symbols. 1. Multiply the whole numbers. 2. Multiply the fraction. 3. Add.

           1. $250,000 \times 6 =$ 1,500,000

           2. $250,000 \times \frac{1}{2} =$ + 125,000

           3.   1,625,000

ANSWER: True

Comprehension Questions

Values on the graph can be compared and conclusions can be drawn.

EXAMPLE   Between the American Federation of Teachers (AFT) and the Communication Workers of America what is the difference in the number of members?

    a. 250,000 workers   **c.** 650,000 workers   **e.** 100,000 workers
    b. 400,000 workers   **d.** 750,000 workers

       STEP 1   Find the name "American Federation of Teachers." Calculate this union's membership total by multiplying the number of symbols by the key's value.
               $6 \times 250,000 = 1,500,000$

       STEP 2   Find the name "Communication Workers of America." Calculate this union's membership by multiplying the number of symbols by the key's value.
               $3 \times 250,000 = 750,000$

       STEP 3   Subtract to find the difference between the memberships of the two unions.
               $1,500,00 - 750,000 = 750,000$

ANSWER: d. 750,000 workers

Practice Graph I

Graph I, read from left to right across the page, is an example of a horizontal pictograph. Graph I shows the hourly earnings of jobs in particular goods-producing and service-producing industries as of January 2010.

GRAPH I

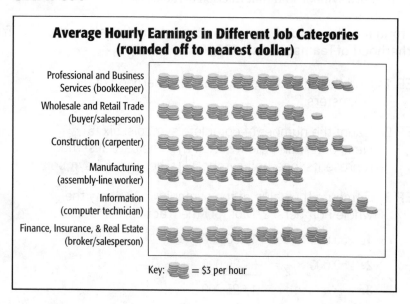

Scanning Graph I

Fill in each blank as indicated.

1. Find each of the following:

   a. Graph Title: _____

   b. Vertical Axis Names:

   _____

   _____

   _____

   _____

   _____

2. Each ![coin] symbol is equal to _____ in earnings per hour.

3. The hourly pay for a carpenter is found by looking at the job category called _____.

Reading Graph I

Decide whether each statement is true or false and circle your answer.

4. The average hourly pay for a construction worker is $14 per hour.   True   False

5. A computer technician can expect to earn average hourly wages of approximately $26.00 per hour.   True   False

6. Graph I shows that salespeople working in clothing stores usually earn an average of $15.00 per hour.   True   False

Comprehension Questions

Answer each question by choosing the best multiple-choice response.

7. Hourly earnings in the information trades are generally ＿＿＿＿＿＿＿ other occupations.

   a. higher than   **d.** twice that of
   b. lower than   **e.** none of the above
   c. the same as

8. What is the difference between the hourly wages of an assembly line worker and the earnings of a carpenter?

   a. $1.50   **d.** $5.00
   b. $2.00   **e.** $6.00
   c. $3.50

9. Which job category shown in Graph I has the lowest hourly earnings?

   a. retail trade   **d.** construction
   b. manufacturing   **e.** finance
   c. transportation

10. Which statement best describes Graph I?

   a. More physically difficult jobs pay higher hourly wages.
   b. Workers skilled in a trade or a craft make lower wages than workers skilled in management or supervision.
   c. Average hourly earnings for union workers range from $2 to $3 per hour higher than for non-union workers.
   d. Average hourly earnings for workers in the job categories shown range from $18.00 to $26.00 per hour.
   e. Average hourly earnings for non-union workers are increasing at a rate faster than earnings for union workers.

Practice Graph II

Graph II is a vertical pictograph. Symbols are shown in vertical columns. A vertical pictograph is often used to compare information about one item over a period of time. Graph II shows the change in the unemployment rate over several years.

GRAPH II

Source: U.S. Department of Labor

Scanning Graph II

Fill in each blank as indicated.

1. The information presented on Graph II was obtained from the source known as the _____.

2. The unemployment rates of the labor force are given from the year _____ to the year _____.

3. Graph II tells about the unemployment rates as a _____ of the civilian labor force.

Reading Graph II

Decide whether each sentence is true or false and circle your answer.

4. The highest unemployment rate shown occurred in 2008.   True   False

5. In 2007, the unemployment rate was about 4.5%.   True   False

6. According to Graph II, the lowest unemployment rate occurred in 2005.   True   False

Comprehension Questions

Answer each question by choosing the best multiple-choice response.

7. What is the difference in the unemployment rate between 2005 and 2008?

    a. 1%
    b. 2%
    c. 3%
    d. 4%
    e. 5%

8. The greatest unemployment rate decrease occurred between what years?

    a. 2006–2007
    b. 2004–2005
    c. 2007–2008
    d. 2008–2009

9. From Graph II, you could assume that the greatest percentage of the workforce was unemployed in what year?

    a. 2007
    b. 2005
    c. 2008
    d. 2009
    e. none of the above

10. Which statement best describes Graph II?

    a. Unemployment rates rose steadily from 2005–2007.
    b. Unemployment rates dropped steadily from 2008–2009.
    c. Unemployment rates continuously increased from 2007 to 2009.
    d. Unemployment rates follow a consistent pattern of increasing one year and then decreasing the next.
    e. Unemployment rates were higher in 2007 than in any other year since 2004.

Practice Graph III

For the purpose of making comparisons, some pictographs, such as Graph III below, are used to display more than one type of information.

GRAPH III

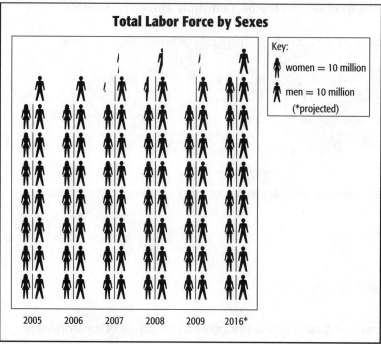

Source: U.S. Department of Labor

Scanning Graph III

Fill in each blank as indicated.

1. In Graph III, the asterisk (*) means that the information for 2016 is
   _____.

2. Each ♀ or ♂ on the graph represents _____ workers.
                                              (number)

3. Graph III is a labor force comparison between _____ and
   _____.

Reading Graph III

Decide whether each sentence is true or false and circle your answer.

4. The year with the greatest number of women in the work force was 2006.   True   False

5. During 2008, women made up more than one-half of the *total* labor force. True False

6. By 2006 the *total* labor force was approximately 150 million people. True False

Comprehension Questions

Answer each question by choosing the best multiple-choice response.

7. In 2016 the number of women in the work force will be _____ the number of men in the work force in 2006.

   a. equal to
   b. twice
   c. triple
   d. four times
   e. one-half

8. Between 2006 and 2008, the female labor force increased by what amount?

   a. 85 million
   b. 55 million
   c. 5 million
   d. 100 million
   e. 10 million

9. Which diagram best represents a relationship shown on Graph III?

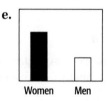

10. Which statement is true according to Graph III?

   a. In comparison to 2005, the number of people in the total labor force is expected to decline in the 21st century.
   b. Men have been and are expected to continue as the larger of the two groups of employed workers.
   c. Women are expected to become the larger group of employed workers by the first part of the 21st century.

Pictographs: Applying Your Skills

When workers are unemployed, their families undergo many hardships. Some people may be able to find other work when they are fired or laid off. Others may not be so fortunate and must rely on unemployment compensation and public assistance for a while. One way to guard against becoming unemployed is to develop skills that will be in demand in the future and continually seek training and education.

GRAPH A

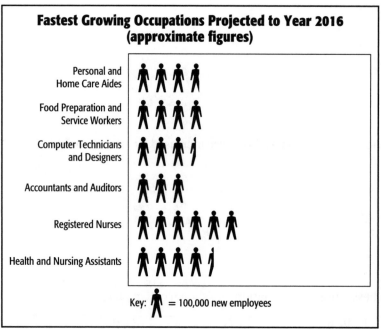

Source: U.S. Bureau of Labor Statistics

Answer each question by filling in the blank, answering true or false, or choosing the best multiple-choice response.

1. Graph A shows the fastest growing occupations to the year _____.

2. The total number of people projected to be employed as health and nursing assistants by 2016 will be approximately _____.

3. The number of registered nurses is expected to be twice the number of accountants and auditors   True   False

4. According to Graph A, the number of computer technicians in 2016 will grow by 400,000.   True   False

5. The information shown in Graph A indicates that more people are   True   False
projected to be employed in hospitals and nursing homes than in restaurants.

6. What is the approximate number of people projected for employment as food
preparers and service workers?

    a. 600,000
    b. 800,000
    c. 700,000
    d. 400,000
    e. 650,000

7. According to Graph A, the difference in the number of people to be employed as
_____ will be double the amount for accountants.

    a. nurses
    b. medical assistants
    c. teachers
    d. physical and occupational therapists
    e. engineers and system analysts

8. If the number of people employed as food preparers and service workers
increases by 50% more than the graph predicts, what will be the total number
of workers in 2016?

    a. 100,000
    b. 200,000
    c. 400,000
    d. 600,000
    e. 800,000

9. Which diagram best represents a relationship shown on Graph A?

a.   **b.**   **c.** 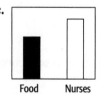   **d.** 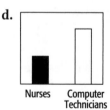   **e.**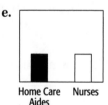

Nurses Accountants   Computer Technicians Home Care Aides   Food Preparers Nurses   Nurses Computer Technicians   Home Care Aides Nurses

10. All of the following conclusions can be drawn from Graph A EXCEPT:

    a. By the year 2016, more people will be employed as computer specialists than
as accountants.
    b. By the year 2016, the increase in the number of personal and home care aides
will be smaller than the increase in the number of nurses.
    c. By the year 2016, there will be over 1.4 million new jobs in all of the health-related
occupations.

In an effort to establish fair working conditions for all, employees joined together to form unions in the mid-1800s. This alliance enabled the common worker to have a say in how industries were run.

Since their inception, unions have grown to incorporate many fields of labor. Workers join with others in their profession, enabling them to address issues important to their specific industry. The United Auto Workers union has a much different agenda than the teacher's union, for example.

GRAPH B

U.S. Membership in Some AFL-CIO Affiliated Unions: 2010

Plumbers, Pipefitters	▢▢▢▢▢▢▢
Painters and Allied Professionals	▢▢▢
Teachers	▢▢▢▢
Government Employees	▢▢▢▢▢▢▢▢▢▢
Postal Workers	▢▢▢▢▢

Key: ▢ = 50,000 members

Source: Industrial Relations Data and Information Services

Answer each question by filling in the blank, answering true or false, or choosing the best multiple-choice response.

1. Graph B represents the U.S. _____ in some AFL-CIO affiliated unions.

2. The union with the smallest membership as represented on Graph B is

   _____.

3. There are twice as many members in the government employees union as in the postal workers' union.   True   False

4. The painters' union has a larger membership than the teachers' union.   True   False

5. Graph B shows that of all the unions listed, government employees have the largest membership.   True   False

6. Which unions, when combined, have slightly under 500,000 members?

     a. plumbers and teachers
     b. government workers, painters, and plumbers
     c. painters and plumbers
     d. painters, postal workers, and plumbers
     e. postal workers and painters

7. If the number of members in the plumbers and pipefitters' union increased by 10%, how many members would this union have?

     a. 357,500
     b. 410,000
     c. 303,000
     d. 270,000
     e. 40,000

8. If the number of members in the _____ union were to decrease by 50%, this union would have only 150,000 members.

     a. government employees
     b. teachers
     c. plumbers and pipefitters
     d. postal workers
     e. painters

9. From Graph B, how many more members are in the plumbers' union than in the painters' union?

     a. 100,000
     b. 400,000
     c. 175,000
     d. 600,000
     e. 300,000

10. Which statement is true according to Graph B?

     a. There are four times more members in the government employees' union than in the painters' union.
     b. There are more than twice as many members in the government employees' union than in the plumbers and pipefitters' union.
     c. 50% of all electrical workers belong to a union.
     d. In 2010 there were more members in AFL-CIO affiliated unions than ever before.
     e. Membership in unions is declining overall.

Circle Graphs

A **circle graph,** or **pie graph,** shows an entire quantity divided into various parts. Each part of the circle is called a **segment** and has its own name and value. In most cases, values on circle graphs consist of either parts, a fraction, or a percent of a whole.

Circle graphs often illustrate budgets and expenses. Graph A shows the sources for each dollar that the federal government receives.

GRAPH A

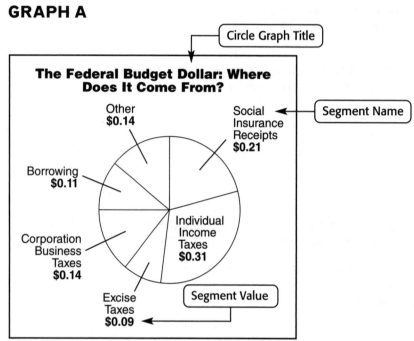

Circle Graph Title

The Federal Budget Dollar: Where Does It Come From?

Other $0.14

Social Insurance Receipts $0.21 — Segment Name

Borrowing $0.11

Individual Income Taxes $0.31

Corporation Business Taxes $0.14

Excise Taxes $0.09 — Segment Value

Source: U.S. Office of Management and Budget

To answer questions about a circle graph, follow the directions below.

Scanning the Graph

To scan a circle graph, read the graph title and the segment names.

EXAMPLE   Graph A shows that part of the federal budget dollar comes from individual income _____.

STEP 1   Find and read the graph title, "The Federal Budget Dollar: Where Does It Come From?" The title is important but does not give the answer.

STEP 2   Read the segment names. Read clockwise until you find the segment name, "Individual Income Taxes."

ANSWER: **taxes**

Reading the Graph

To read a circle graph, locate the name and value of each segment. Each segment value can represent a fraction, percent, or number.

EXAMPLE   From every dollar the government receives, 21¢ comes from   True   False
social insurance receipts.

STEP 1   Read clockwise on the graph until you find the
segment named "Social Insurance Receipts."

STEP 2   Near the segment name, read the dollar value. The
value for social insurance receipts is $0.21, which is
another way to write 21¢.

ANSWER: **True**

Comprehension Questions

By comparing segment names and segment values, you can draw conclusions from the graph.

EXAMPLE   For each dollar, what is the difference between the amount
that the government receives from borrowing and from excise
taxes?

a. 1¢
b. 2¢
c. 3¢
d. 20¢
e. 15¢

STEP 1   Find the segment name and value "Borrowing—11¢."

STEP 2   Find the segment name and value "Excise taxes—9¢."

STEP 3   Find the difference between the two values by
subtracting.

11¢ − 9¢ = 2¢

ANSWER: **b. 2¢**

Practice Graph I

Graph I is a circle graph that shows where each dollar budgeted by the federal government is spent. Each pie-shaped segment represents a part of a dollar ($1.00).

GRAPH I

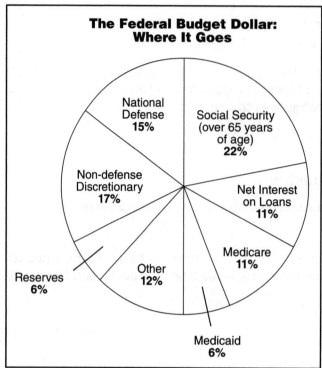

The Federal Budget Dollar: Where It Goes

National Defense 15%

Social Security (over 65 years of age) 22%

Non-defense Discretionary 17%

Net Interest on Loans 11%

Medicare 11%

Reserves 6%

Other 12%

Medicaid 6%

Source: U.S. Office of Management and Budget

Scanning Graph I

Fill in each blank as indicated.

1. Find each of the following:

   a. This graph is about _____ .

   b. The main categories of the federal budget are

   _____ _____

   _____ _____

   _____ _____

   _____ _____

2. Payments made to citizens over 65 come from _____ .

3. The category to which Congress allocates money for missiles, tanks, and military staff pay is _____ .

Reading Graph I

Decide whether each statement is true or false and circle your answer.

4. At least one-half of each federally budgeted dollar is used for Medicaid and Medicare health payments.

    True   False

5. Of each dollar the government spends, 25% is spent on national defense.

    True   False

6. The smallest amount spent in any one category is in non-defense expenditures.

    True   False

Comprehension Questions

Answer each question by choosing the best multiple-choice response.

7. From each dollar collected, the government spends _____ toward net interest on loans.

    a. 1 cent
    b. 6 cents
    c. 11 cents
    d. 17 cents
    e. 22 cents

8. The difference between money spent for national defense and money paid in Social Security benefits to individuals is _____ for each dollar.

    a. 14¢
    b. 6¢
    c. 11¢
    d. 8¢
    e. 7¢

9. Approximately, how much of the budget is spent on areas *other than* Social Security and Medicare?

    a. $\frac{1}{3}$
    b. $\frac{1}{4}$
    c. $\frac{2}{3}$
    d. $\frac{3}{4}$
    e. $\frac{4}{5}$

10. Which statement summarizes Graph I?

    a. The federal dollar is equally divided among several categories.
    b. The federal government collects most of the money for its budget from personal income taxes.
    c. The two leading government expenditures are for Social Security payments and non-defense Discretionary.
    d. The interest paid on the national debt increased 10% between 1989 and 1990.

Practice Graph II

Graph II shows the number of deaths caused by motor vehicles in 2007. Motor vehicle crashes are the leading cause of death among Americans ages 1 to 34. Each segment of the circle represents the percentage of people killed in each category of motor vehicle accidents (or as a pedestrian killed by a motor vehicle).

GRAPH II

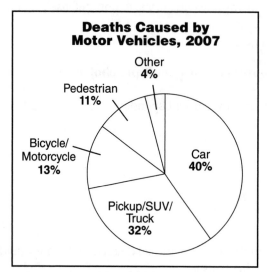

Source: U.S. National Highway Traffic Safety Administration, 2009

Scanning Graph II

Fill in each blank as indicated.

1. Graph II shows the percent of deaths caused by _____.

2. From Graph II, a person walking in a crosswalk who gets hit by a car would be included as a death in the category of _____.

3. The causes of death are divided among _____ categories.
   (number)

Reading Graph II

Decide whether each sentence is true or false and circle your answer.

4. 47% of all deaths in motor vehicles are caused by car accidents.   True   False

5. Sport utility vehicles and trucks show an increase in causes of deaths.   True   False

6. From Graph II, slightly less than three-fourths of all deaths are caused by trucks and cars.   True   False

Comprehension Questions

Answer each question by choosing the best multiple-choice response.

7. Deaths caused by pickups/SUVs/trucks exceed pedestrian and bicycle deaths by _____ percent.

    a. 24
    b. 10
    c. 19
    d. 8
    e. 14

8. If the percentage of pedestrian deaths doubles in the next year, what will be the new allocation for deaths in this category?

    a. 8%
    b. 14%
    c. 22%
    d. 5%
    e. 17%

9. From Graph II, what causes approximately one-third of all motor vehicle deaths?

    a. Car
    b. Pickup/SUV/Truck
    c. Motorcycle/Bicycle
    d. Other
    e. Pedestrian

10. All of the following statements can be inferred from Graph II EXCEPT:

    a. More persons die in cars than any other vehicle type.
    b. The majority of deaths are caused by SUVs, trucks, and cars.
    c. Approximately three-fourths of deaths in motorized vehicles are due to SUV, truck, and car accidents.
    d. The majority of deaths to pedestrians are caused by a car or truck.
    e. Disregarding the "Other" category, pedestrian accidents total the fewest deaths caused by motor vehicles.

Practice Graph III

In some cases, more than one graph can be used to make comparisons. The following circle graphs show how the federal budget was estimated to change during three consecutive years.

GRAPH III

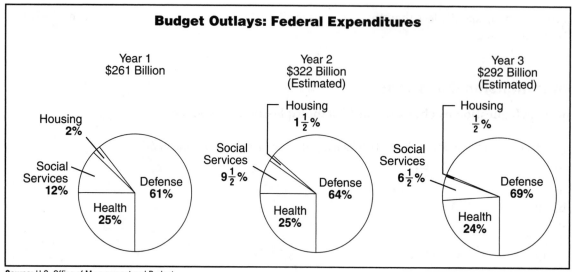

Budget Outlays: Federal Expenditures

Source: U.S. Office of Management and Budget

Scanning Graph III

Fill in each blank as indicated.

1. The amount of federal expenditures shown for Year 2 and Year 3 are _____ figures.

2. The total budget for Year 2 and Year 3 is estimated to be $_____ billion and $_____ billion, respectively.

3. According to Graph III, the federal government's budget is divided into these four main categories:

   _____

   _____

   _____

   _____

Reading Graph III

Decide whether each sentence is true or false and circle your answer.

4. In Year 1, 2% of federal expenditures was for housing.   True   False

5. The percent of the federal budget allocated for defense is expected to
   decrease over the 3-year period.   True   False

6. The budget outlay for defense in Year 3 is estimated to be 69%.   True   False

Comprehension Questions

Answer each question by choosing the best multiple-choice response.

7. From Year 1 to Year 3, the percent allocated for which item will change the least?

   a. health
   b. housing
   c. social services
   d. defense

8. What budget item was projected to be reduced more than any other by Year 3?

   a. defense
   b. housing
   c. social services
   d. health

9. Which diagram best shows the total budget outlays?

  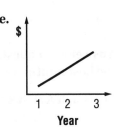

10. Which statement best reflects the information in Graph III?

   a. For Year 1 to Year 3, there is a projected increase of total federal expenditures
   in all categories.
   b. For Year 1 to Year 3, there is a projected increase in the percentage
   of total federal expenditures going for defense.
   c. For Year 2, total federal expenditures were projected to be less than
   they were in Year 1 for both housing and health.
   d. Unemployment will increase in Year 3 because of decreasing federal
   expenditures for social services.

Circle Graphs: Applying Your Skills

The increasing incidence of breast and other related cancers in women has become a major area of research by the medical profession. Medical centers, hospitals, private and public agencies and organizations are currently involved in researching cures of all types of cancers.

PROJECTIONS OF CANCER IN MEN TO RISE BY 30% BY 2015

Equally disturbing are the startling increases in prostate and other related cancers in men, especially in the United States. Medical professionals and cancer research teams are increasingly shifting their attention to cancer in males.

GRAPH A

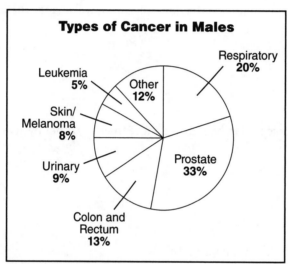

Types of Cancer in Males

Respiratory 20%
Leukemia 5%
Other 12%
Skin/Melanoma 8%
Urinary 9%
Prostate 33%
Colon and Rectum 13%

Source: Center for Disease Control, 2010

Answer each question by filling in the blank, answering true or false, or choosing the best multiple-choice response.

1. Graph A shows the percentage of males who suffer from a variety of
   _____.

2. The highest percentage of men suffer from _____ cancer.

3. During the past 10 years, prostate cancer has increased at the fastest rate of all types.　　　True　　　False

4. Graph A shows prostate and respiratory cancers combined account for the majority of types of cancer in men.　　　True　　　False

5. The percentage of urinary, colon, and skin cancers combined in males is equal to the percentage of respiratory cancers.　　　True　　　False

6. One out of every _____ men diagnosed with cancer suffer from prostate cancer.

    a. 2
    b. 9
    c. 3
    d. 5
    e. 6

7. What percent of all males diagnosed with cancer suffer from either urinary or skin cancer?

    a. 33%
    b. 22%
    c. 19%
    d. 20%
    e. none of the above

8. Approximately two out of _____ men suffer from respiratory cancers.

    a. ten
    b. five
    c. three
    d. thirteen
    e. nine

9. Which diagram best illustrates the relationship between the percentage of males with prostate cancer and the percentage of males with leukemia in Graph A?

a.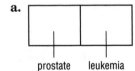
b.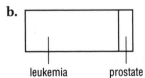
c.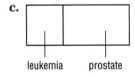
d.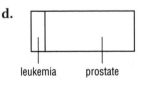

10. Graph A is best summarized as follows:

    a. The leading causes of cancer in males are too much exposure to sun and toxic materials.
    b. The number of cases of prostate cancer is greater than the number of cases of other types of cancer in males.
    c. Having annual medical check-ups is the only way to reduce prostate cancer.

Graph B shows two circle graphs used to compare the changes in civilian labor forces projected to the year 2016.

Because of changes in work, education, and mobility, individuals are tending to move where the jobs are and where opportunity arises. For many individuals and their families, the United States offers opportunities that don't exist in other parts of the world.

GRAPH B

U.S. Civilian Labor Force

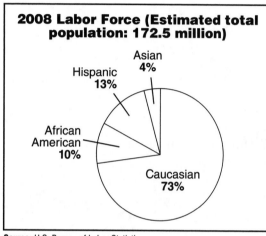

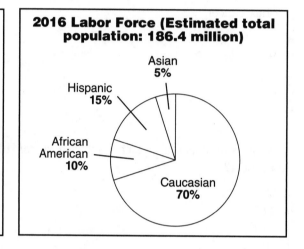

Source: U.S. Bureau of Labor Statistics

Answer each question by filling in the blank, answering true or false, or choosing the best multiple-choice response.

1. The graphs above show the U.S. _____ _____ for 2008 and 2016.

2. A more appropriate title for the two graphs above could be
   a. U.S. Civilian and Army Labor Force
   b. U.S. Civilian Labor Force by Ethnic Origin
   c. U.S. Labor Force by Sex, Race, and Ethnicity
   d. Percentage of Changes in Total Labor Force

3. Graph B shows that the African American labor force is projected to remain the same.   True   False

4. There are about twice as many Hispanics projected to be in the U.S. labor force in 2016 as compared to 2008.   True   False

5. From the graphs, you can infer that more minorities are projected to enter the labor force.   True   False

6. The majority of the civilian labor force consists of which group?

   a. African Americans
   b. Caucasians
   c. Hispanics
   d. Asians

7. The minority civilian labor force is projected to increase from 27% in 2008 to _____% in 2016.

   a. 28
   b. 23
   c. 30
   d. 35
   e. 72

8. From 2008 to 2016, which group is projected to have the greatest increase in U.S. civilian labor force?

   a. African Americans
   b. Caucasians
   c. Asians
   d. Hispanics

9. Which diagram best compares the percent of changes in ethnic origin from 2008 to 2016?

a.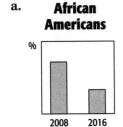
b.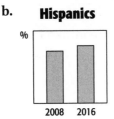
c.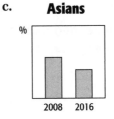
d.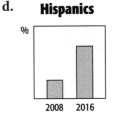
e.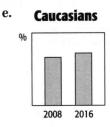

10. All the statements below can be concluded from Graph B EXCEPT:

   a. There is an increase in minorities across the U.S. labor force.
   b. The percent of minority workers is smaller in 2008 than projected in 2016.
   c. Caucasians comprise the largest percentage of the U.S. civilian labor force.
   d. More individuals are entering the labor force than in the past.
   e. Less than 25% of the U.S. civilian labor force in 2008 were minorities.

Bar Graphs

Bar graphs get their name from the thick bars with which they are drawn. They are generally more complicated to read than pictographs or circle graphs.

Bar graphs contain labeled points (usually names or numbers) along the **horizontal axis** (often called the "*x*" axis) and **vertical axis** (often called the "*y*" axis). The values represented by an individual bar, called an **information bar,** are determined by both its height (or length) and by its position along an axis. Follow the steps outlined below to learn how to accurately read bar graphs.

GRAPH A

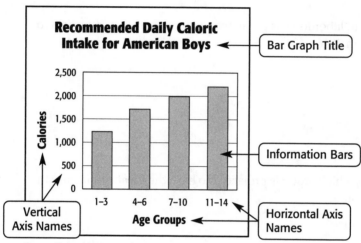

Source: National Academy of Sciences

To answer questions about bar graphs, follow the sequence below.

Scanning the Graph

To scan a bar graph, find the graph title and the names of the axes.

EXAMPLE   Graph A shows the recommended daily calorie intake for American boys whose ages range from 1 to _____ years.

STEP 1   Find the horizontal axis name Age Groups.

STEP 2   Scan from left to right, reading the labeled points at the bottom of each information bar. The bar farthest to the right represents the oldest age group, 11–14.

ANSWER: **14**

Reading the Graph

To read a bar graph, find the values represented by the information bar. Read these values as labeled points along the horizontal and vertical axes.

EXAMPLE The amount of daily calories recommended for a boy 12 years of age is 2,200.   True   False

STEP 1 Find the horizontal axis name, "11–14" years old. This axis name includes those boys whose ages are 11, 12, 13, and 14.

STEP 2 Move your eyes from the bottom to the top of this information bar. After doing this, move across the graph to the left until you reach the vertical axis. You might use the edge of a paper to make sure you are aligning accurately.

STEP 3 Read the vertical axis name and number, "Calories, 2,200." The number of calories for a boy 12 years old is 2,200.

ANSWER: True

Comprehension Questions

Comprehension questions involve comparing values represented by several information bars and then drawing conclusions. Also, bar graphs are useful in showing trends. A **trend** is a pattern that can be seen from the information contained on the graph. From a trend, it is often possible to make predictions about future occurrences.

EXAMPLE As boys get older, the number of calories consumed daily should

a. stay the same   **c.** decrease
b. increase   **d.** level off

STEP 1 Find the horizontal axis names that refer to the specific age groups. Scan from the bottom to the top of each bar and identify the calories for each group.

STEP 2 Compare the amount of calories between the ages of boys from very young to older. For example, ages:

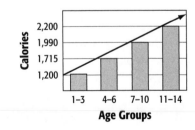

1–3 = 1,230 calories
4–6 = 1,715 calories
7–10 = 1,990 calories
11–14 = 2,200 calories

STEP 3 Draw a conclusion: As boys get older, the number of calories they require increases.

ANSWER: b. increase

Practice Graph I

Graph I shows a vertical bar graph with bars drawn up from the horizontal axis. Also, notice the symbol ⌇ on the vertical axis line. Occasionally, this symbol is used to show that values have been omitted from the axis in order to save space. On the graph below, weights from 1 to 135 have been omitted.

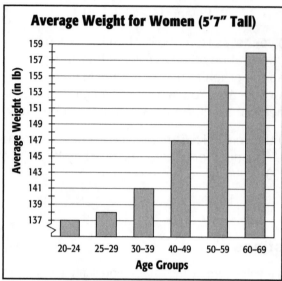

GRAPH I

Source: Society of Actuaries

Scanning Graph I

Fill in each blank as indicated.

1. Find each of the following:

   a. Graph Title: _____

   b. Vertical Axis Title: _____

   c. Horizontal Axis Title: _____

2. The youngest women represented are from _____ to _____ years old.

3. The greatest average weight shown on the graph for women 5′7″ tall is _____ pounds.

Reading Graph I

Decide whether each statement is true or false and circle your answer.

4. The average weight for a 40-year-old woman is 147 pounds.   True   False

5. A 152-pound woman in the age group 30–39 years is below the average weight.   True   False

6. The highest average weight occurs for women in the 50–59 year age group.   True   False

Comprehension Questions

Answer each question by choosing the best multiple-choice response.

7. On the average, as women get older, the trend is for their weight to

   a. decrease
   b. stay the same
   c. increase
   d. increase until age 30
   e. none of the above

8. What is the difference in average weight between a 69-year-old woman and a 29-year-old woman?

   a. 12 pounds
   b. 10 pounds
   c. 20 pounds
   d. 18 pounds
   e. 35 pounds

9. From Graph I, you can tell that the greatest gain in weight occurs from age group _____ to _____.

   a. 20–24, 25–29
   b. 25–29, 30–39
   c. 30–39, 40–49
   d. 40–49, 50–59
   e. 50–59, 60–69

10. Graph I tells about

   a. the average weight and height for 5-foot 7-inch-tall adults
   b. average weights for all women between the ages of 20 and 69 years
   c. average weight at specific ages for women 5 feet 7 inches tall
   d. average weights for obese women

Practice Graph II

Graph II is a horizontal bar graph. The information bars are drawn across the graph from left to right. Although they are not as common as vertical bar graphs, it is useful to know how to read horizontal bar graphs.

GRAPH II

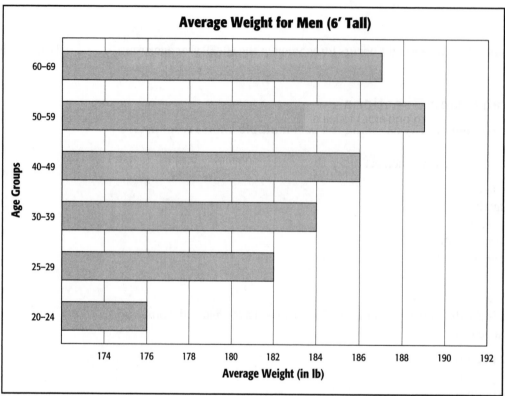

Source: Society of Actuaries

Scanning Graph II

Fill in each blank as indicated.

1. Graph II represents the _____ for men 6 feet tall.

2. The highest weight that could be represented on this graph is _____.

3. The total age span represented by Graph II starts at age _____ and extends to age _____.

Reading Graph II

Decide whether each sentence is true or false and circle your answer.

4. The average weight for a 6-foot-tall, 28-year-old man is 184 pounds.   True   False

5. The graph shows that the lowest average weight for 6-foot-tall men occurs in the 20 to 24 year age group.   True   False

6. A 6-foot-tall man in the 50- to 59-year age group may weigh 189 pounds.   True   False

Comprehension Questions

Answer each question by choosing the best multiple-choice response.

7. On the average, a man's weight can be expected to increase until what age?

   a. 45
   b. 49
   c. 39
   d. 63
   e. 60

8. The greatest average weight increase shown on the graph occurs from the ages of

   a. 20–24 to 25–29
   b. 25–29 to 30–39
   c. 30–39 to 40–49
   d. 40–49 to 50–59
   e. 50–59 to 60–69

9. Which graph best shows the weight trend for men 6 feet tall between the ages of 20 and 69?

   a. 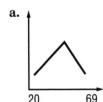   b. 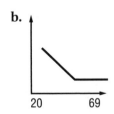   c. 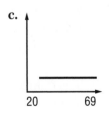   d. 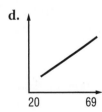   e.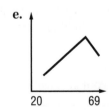

10. Which statement best describes the information on Graph II?

    a. Men tend to gain weight as their age increases.
    b. On the average, men 6 feet tall gain weight up to age 59, and tend to lose weight after age 60.
    c. Taller men gain more weight in their later years than shorter men.
    d. Men tend to gain weight during retirement years.
    e. Weight gain is a problem for most people.

Practice Graph III

Graph III is a double bar graph. This graph contains two kinds of information on the same graph. It shows the average weight for both men and women who are 5 feet 5 inches tall.

In each age group, a bar representing women's weight stands next to a bar representing men's weight. The bars are identified by a key in the upper left hand corner of the graph. Graphs containing more than one bar are used to make comparisons.

GRAPH III

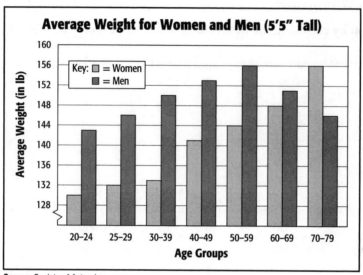

Source: Society of Actuaries

Scanning Graph III

Fill in each blank as indicated.

1. In Graph III, the dark grey bar represents the average weight for _____, and the light grey bar represents the average weight for _____.

2. The youngest age group represented on Graph III is from age _____ to _____, and the oldest age group is from _____ to _____.

3. The highest weight that could be represented on this graph is _____.

Reading Graph III

Decide whether each sentence is true or false and circle your answer.

4. At age 53, a 5'5" man is likely to weigh about 12 pounds more than a 5'5" woman of the same age.

   True   False

5. When both are in their seventies, 5'5" men are usually heavier than 5'5" women.

   True   False

6. The average weights of 5'5" men and women are the closest to being equal between ages 60 and 69 years.

   True   False

Comprehension Questions

Answer each question by choosing the best multiple-choice response.

7. From 30 to 39 years, 5'5" women weigh about _____ pounds less than 5'5" men.

   a. 4
   b. 7
   c. 10
   d. 17
   e. 20

8. The trend on Graph III indicates that as a woman gets older, she is expected to _____ weight.

   a. gain
   b. lose
   c. gain and then lose
   d. lose and then gain

9. Women who are 5'5" tall and between the ages of 70–79 are the closest in weight to men 5'5" tall and between what ages?

   a. 20–24
   b. 25–29
   c. 40–49
   d. 50–59
   e. 60–69

10. Which statement best describes the information on Graph III?

   a. In the age span shown, the pattern of weight gain for men is the same as the pattern of weight gain for women.
   b. Through their fifties, taller men and women are heavier than shorter men and women.
   c. Both men's and women's weight are affected by retirement.
   d. Both men and women tend to gain weight through their fifties, but men, unlike women, tend to lose weight in later years.

Bar Graphs: Applying Your Skills

Americans are becoming more concerned with the quality of life and with their health. Weight and disease control, nutrition, and physical fitness are popular topics. Additionally, our society is placing great importance on improving health care services while controlling costs.

Graph A shows the sources of health care payments.

GRAPH A

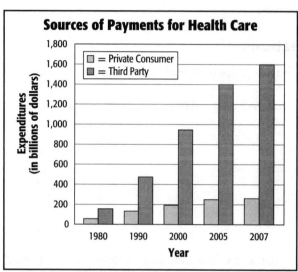

Source: Health Care Financing Administration

Answer each question by filling in the blank, answering true or false, or choosing the best multiple-choice response.

1. The sources of payments for health care shown on the graph are _____ and _____.

2. From 1980 to 2007, third party expenditures increased by approximately _____ dollars.

3. Private consumer payments were about half the amount of third party payments in 2000.   True   False

4. From 2005 to 2007, private consumer payments remained about the same amount.   True   False

5. In 2005, third party health care payments were approximately 300 billion dollars   True   False
greater than private consumer payments.

6. According to Graph A, the smallest increase in third party payments occurred
from

    a. 1980 to 1990   **c.** 2000 to 2005
    b. 1990 to 2000   **d.** 2005 to 2007

7. What was the approximate total health care expenditures paid by both third
party and private consumer sources in 2000?

    a. $2 trillion, 350 billion   **d.** $1 trillion, 100 billion
    b. $1 trillion, 150 billion   **e.** $900 billion
    c. $2 trillion, 800 billion

8. From 2005 to 2007, total expenditures by third parties and private consumers
rose by approximately what amount?

    a. $10 billion   **c.** $200 billion
    b. $50 billion   **d.** $150 billion

9. According to Graph A, private consumers paid how many more dollars for health
care in 2000 than in 1980?

    a. $70 billion   **d.** $50 billion
    b. $90 billion   **e.** $200 billion
    c. $130 billion

10. Which statement best describes Graph A?

    a. Health care expenses are at an all time low.
    b. Private consumers should pay more of the cost of health care.
    c. Third party payments rose steadily between 1990 and 2007, while
        private consumer payments remained about the same.
    d. Expenditures by third parties and private consumers rose steadily
        between 1990 and 2005.
    e. Expenditures by private consumers have consistently been greater
        than expenditures by third party sources.

RAISE DRIVING AGE TO 18?

According to some public health researchers, accidents are the fourth most likely cause of death, following heart attacks, cancer, and strokes. Raising the legal age for driving to age 18 could save at least 2,000 lives a year.

GRAPH B

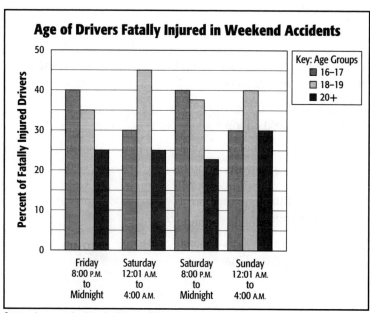

Source: Insurance Institute for Highway Safety

Answer each question by filling in the blank, answering true or false, or choosing the best multiple-choice response.

1. Graph B represents drivers who are _____ in weekend car accidents.

2. Teenagers 16 and 17 years of age account for _____ percent of driver deaths on Saturday nights between 8:00 P.M. and midnight.

3. 16- and 17-year-olds make up the same percent of drivers who are fatally injured   True   False
   from 8:00 P.M. to midnight Friday and from 8:00 P.M. to midnight Saturday.

4. After midnight Saturday, more 16- and 17-year-old drivers have accidents than   True   False
   do drivers in the other two age groups.

5. 16- and 17-year-old drivers comprise the same percentage of fatal accidents as 20+ year-old drivers on

    a. Friday 8 P.M. to midnight
    b. Saturday 12:01 A.M. to 4:00 A.M.
    c. Saturday 8 P.M. to midnight
    d. Sunday 12:01 A.M. to 4:00 A.M.

6. The highest percentage of fatally injured drivers, 20 years and older, is _____ and occurs on Sunday 12:01 A.M. to 4:00 A.M.

    a. 55%
    b. 30%
    c. 15%
    d. 25%
    e. 32%

7. The time period when the highest percentage of fatally injured drivers is in the 18- to 19-year-old driving group is

    a. Friday 8 P.M. to midnight
    b. Saturday 12:01 A.M. to 4:00 A.M.
    c. Saturday 8 P.M. to midnight
    d. Sunday 12:01 A.M. to 4:00 A.M.

8. Of the groups listed, which has an average accident rate of 35% for the entire weekend?

    a. 16–17 year olds
    b. 18–19 year olds
    c. 20+ year olds

9. Which statement best describes Graph B?

    a. Fatally injured drivers are most often 16 to 17 years old for every time period during the weekend.
    b. Teenagers always have more fatal accidents than adults.
    c. On Friday night between 8:00 P.M. and midnight, 85% of all drivers who are fatally injured are 18 to 19 years old.
    d. Of the three age groups, drivers who are 20 years old and older are the least likely to be fatally injured in a weekend car accident.

Line Graphs

Line graphs are especially useful in showing trends and developments over a period of time. A line graph can be drawn with either curved or straight lines that extend across the graph in a horizontal direction.

Graph A shows how the number of people living at or below poverty levels changed over a period of several years.

GRAPH A

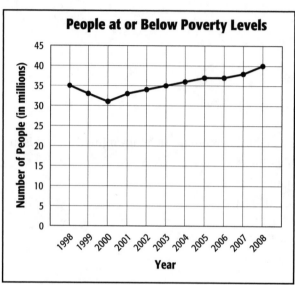

Source: U.S. Census Bureau

To answer questions about a line graph, follow the steps below.

Scanning the Graph

To scan a line graph, find the graph title, the axes names, and the labeled points along each axis.

EXAMPLE   Graph A shows the number of people living below poverty level from 1998 to _____.

STEP 1   Find the horizontal axis title: Year.

STEP 2   Find the last year included on the graph.

ANSWER: **2008**

Reading the Graph

To read a line graph, find the information line, and read the labeled points along the horizontal and vertical axes.

EXAMPLE In 1998, 30 million people were living at or below poverty level.   True   False

    STEP 1   Find 1998 on the horizontal axis. From the bottom of the graph, move directly upward to the information line.

    STEP 2   From this point on the information line, move to the left to the labeled point on the vertical axis. You might use the edge of a paper to make the alignment accurately.

    STEP 3   Read the labeled point on the vertical axis—35.

ANSWER: False. In 1998, 35 million (not 30 million) people were living below poverty level.

Comprehension Questions

Inferences and predictions can be made by comparing values represented on the information line.

EXAMPLE According to Graph A, the number of people living in poverty was the highest during which three years?

    a. 1998, 1999, 2000
    b. 2006, 2007, 2008
    c. 2000, 2001, 2002
    d. 2004, 2005, 2006
    e. 2003, 2004, 2005

    STEP 1   Scan across the information line from left to right. Identify the three points on the information line that are higher than any others.

    STEP 2   For each of these three points, move directly downward to the horizontal axis.

    STEP 3   Read the years labeled directly below the designated points on the information line.

ANSWER: b. 2006, 2007, 2008

Practice Graph I

Practice Graph I uses a single line to show the rise and fall of milk prices.

GRAPH I

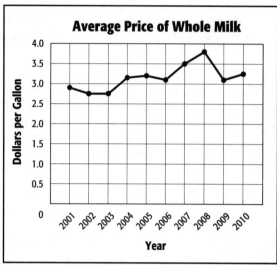

Source: U.S. Bureau of Labor Statistics

Scanning Graph I

Fill in each blank as indicated.

1. Graph I shows the average price of _____.

2. The graph measures the price of milk in _____ per _____.

3. The graph shows the rise and fall of whole milk prices from _____ to _____.

Reading Graph I

Decide whether each statement is true or false and circle your answer.

4. In 2001, a gallon of milk would have cost $3.00.   True   False

5. Between 2006 and 2008, whole milk prices rose.   True   False

6. Two of the years during which decreases in price took place were 2002 and 2006.   True   False

Comprehension Questions

Answer each question by choosing the best multiple-choice response.

7. What is the difference in price between a gallon of milk in 2006 and in 2008?

   a. $0.70
   b. $0.50
   c. $0.90
   d. $0.30
   e. none of the above

8. Between the years 2001 to 2010, what is the difference between the highest and the lowest whole milk prices shown on the graph?

   a. $1.22
   b. $2.50
   c. $1.05
   d. $1.70
   e. $0.70

9. From 2001 to 2006, milk prices

   a. rose steadily
   b. fell steadily
   c. rose sharply, then fell
   d. fell, rose, then fell again
   e. fell sharply, then rose steadily

10. From the information on the graph, you could infer the following:

    a. Farmers went on strike in 1991.
    b. The price of a gallon of whole milk made its greatest price increases during the mid 2000s.
    c. The price of milk rose throughout the 2000s.
    d. Milk prices reflect the rate of customer satisfaction.
    e. none of the above

Practice Graph II

To compare different types of information, a **double line graph** uses more than one line. To prevent confusion, the lines are often drawn differently. A key is used to indicate the meanings of the different lines.

GRAPH II

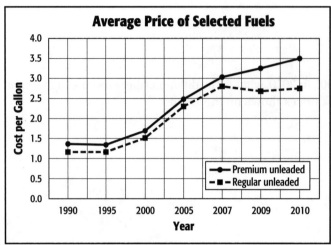

Source: U.S. Energy Information Administration

Scanning Graph II

Fill in each blank as indicated.

1. On Graph II, the solid line represents _____ fuel.

2. The average prices of fuels are measured in _____ per _____.

3. Graph II shows gasoline prices for _____ years.

Reading Graph II

Decide whether each statement is true or false and circle your answer.

4. In 1995 regular unleaded fuel cost an average of $1.34 per gallon.   True   False

5. The price of regular unleaded fuel fell $0.20 per gallon from 2005 to 2008.   True   False

6. In 1990, 10 gallons of premium unleaded would have cost about $13.60.   True   False

Comprehension Questions

Answer each question by choosing the best multiple-choice response.

7. According to the graph, the highest priced fuel has consistently been

   a. regular leaded
   b. regular unleaded
   c. premium unleaded
   d. imported
   e. premium

8. Between 1995 and 2007, the prices of both fuels

   a. stayed the same
   b. rose steadily
   c. tripled
   d. decreased
   e. were cut by half

9. Between _____, the price of regular unleaded increased dramatically.

   a. 1990 and 2000
   b. 2007 and 2010
   c. 1990 and 1995
   d. 1995 and 2007
   e. none of the above

10. The following statement is true according to Graph II:

   a. Fuel prices continued to rise throughout the 1990s.
   b. Fuel prices fell during the 2000s.
   c. Fuel prices fluctuate depending upon supply and demand.
   d. Fuel prices fall most drastically during election years.
   e. Fuel prices show an overall trend of increasing in recent years.

Practice Graph III

Graph III shows the trends in death rates from heart disease across three specific racial groups in the United States. Heart disease has become a popular topic in doctors' offices, hospitals, and homes due to staggering death rates over the years. Currently, advances in medical research have contributed to this trend.

GRAPH III

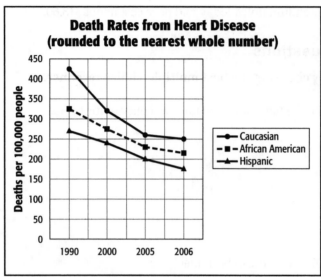

Source: American Heart Organization/Center for Disease Control and Prevention

Scanning Graph III

Fill in each blank as indicated.

1. Graph III represents the death rates from heart disease for _____, _____, and _____.

2. The death rates from heart disease used in Graph III are based on the years between _____ and _____.

3. According to the key, ▲ stands for _____.

Reading Graph III

Decide whether each statement is true or false and circle your answer.

4. In 2006, the death rate from heart disease for Hispanics was 300 per 100,000 people.　　　　True　　False

5. The death rate for African Americans in 2006 was approximately 215 per 100,000 people.　　　　True　　False

6. Death rates from heart disease are greater for Hispanics than for African Americans.　　　　True　　False

Comprehension Questions

Answer each question by choosing the best multiple-choice response.

7. In the year 1990, the death rate from heart disease was greatest among which group?

   a. Caucasian
   b. African American
   c. Hispanic
   d. Asian

8. What was the approximate decrease in the death rate among Caucasians from 1990 to 2006.

   a. 275
   b. 210
   c. 95
   d. 110
   e. 175

9. Between what years was the sharpest decline in death rates from heart disease?

   a. 2000–2005
   b. 1990–2000
   c. 2005–2006
   d. 2000–2006
   e. 2000–2005

10. The following statement is true according to Graph III:

   a. Since 1990, death rates from heart disease have steadily increased.
   b. Hispanics showed a greater decline in heart disease than the other two groups.
   c. There has been a significant decline in death rates from heart disease for all groups represented.
   d. Heart disease has decreased due to changes in medical and health practices.
   e. Hispanics and African Americans are both more likely to die from a heart attack than Caucasians.

Line Graphs: Applying Your Skills

The cost of raising a child has risen greatly during the last several years, and this trend is expected to continue throughout the next century.

COSTS OF RAISING CHILDREN ON THE INCREASE
In the new millennium, costs for health care, food, clothing, shelter, and education are expected to continue to increase. Experts have projected that by the year 2014, it will cost between $200,000 and $300,000 to raise a child to age 17.

GRAPH A

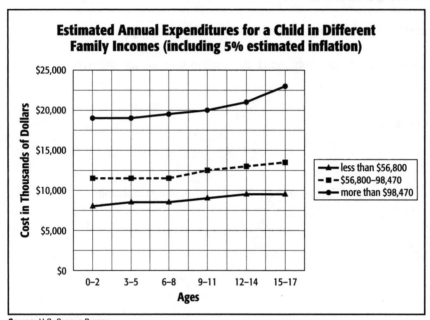

Source: U.S. Census Bureau

Answer each question by filling in the blank, answering true or false, or choosing the best multiple-choice response.

1. Graph A represents the cost of raising a child to age _____.

2. The projected child-raising costs shown on Graph A include an estimated inflation rate of _____.

3. Overall, the yearly cost of raising a child is expected to increase as the child gets older.   True   False

4. The cost of raising a child from the ages of 11 to 17 in a family with an income of $75,000 is more than $100,000.   True   False

5. Graph A shows that the yearly cost of raising a child between the ages of 9–11 is $11,000 for the lowest incomes group.　　　　　True　　　False

6. The period during which the cost of raising a child is the most expensive across all three income groups is from the ages of

　　a. 0–5
　　b. 15–17
　　c. 6–8
　　d. 9–11
　　e. 12–14

7. If 10% of the yearly cost of raising a child is for clothing, the cost of clothing for a child at age 10 in a family with less than $56,800 income will be

　　a. $1,200
　　b. $900
　　c. $1,000
　　d. $1,250
　　e. $1,400

8. According to Graph A, the cost of raising a child at age 13 is approximately _____ more for a family with income of $100,000 than for a family with income of $65,000.

　　a. $15,000
　　b. $3,500
　　c. $8,000
　　d. $4,000
　　e. $12,000

9. If the total cost of raising a child to age 17 is approximately $220,800, the estimated cost of 4 additional years in college, based on the formula of (cost of 1st 17 years) + ($20,000/year/college) would be estimated at

　　a. $342,800　　　　　　**d.** $300,800
　　b. $320,800　　　　　　**e.** $262,890
　　c. $320,800

10. All the following statements can be concluded from Graph A EXCEPT:

　　a. By the time a child is 17 years old, it is expected that he or she could cost at least $170,000 to feed, clothe, and educate, regardless of a family's income.
　　b. As a child gets older, the cost of raising the child gets higher.
　　c. From 0 to 2, the cost of raising a child is relatively constant.
　　d. For a family with income of $56,800 or more, the cost of raising a child in the first 5 years does not increase as rapidly as it does from 12 to 17.
　　e. Inflation accounts for some of the increased costs of raising a child.

SAVE MONEY: WEATHERIZE YOUR HOME

Personal incomes are not increasing at the same rate as utility bills. Therefore, many families are cutting costs by taking steps to insulate and weatherize. The possible savings in weatherizing a home can make the money spent worthwhile.

GRAPH B

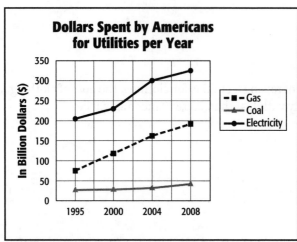

Source: U.S. Census Bureau, 2010

Answer each question by filling in the blank, answering true or false, or choosing the best multiple-choice response.

1. Graph B compares the money spent by Americans for _____, _____, and _____.

2. According to Graph B, Americans spend more money on _____ than on gas or _____.

3. In 2004 the total money spent for the three utilities was approximately $300 billion.   True   False

4. In 2008, Americans spent approximately $385 billion for gas and coal services.   True   False

5. From Graph B, you could conclude that heating costs are rising.   True   False

6. The greatest increase in dollars spent for electricity was from

    a. 2004 to 2008
    b. 2000 to 2004
    c. 1995 to 2000
    d. all the above
    e. none of the above

7. In 2000 _____ more was spent for electricity than for gas.

    a. $110 billion
    b. $120 billion
    c. $95 billion
    d. $130 billion
    e. $80 billion

8. In 2004 the total dollar amount spent for electricity was almost 10 times the amount spent for

    a. coal
    b. telephone
    c. both gas and telephone
    d. all utilities
    e. none of the above

9. Which diagram most accurately describes Graph B?

a.    **b.** 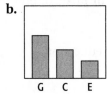   **c.** 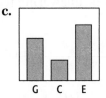   **d.** 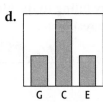   **e.**

10. Which conclusion can be drawn from Graph B?

    a. American consumers are using more utilities every year.
    b. American consumers are spending a larger part of their income for utilities every year.
    c. Prices for utilities have increased more than prices in any other consumer area.
    d. The cost of all utilities has kept up with the rate of inflation.
    e. Less money is spent by American consumers for coal than for other utilities.

Graph Review

Do all the following problems. Work accurately, but do not use outside help. After completing the review, check your answers with the key at the back of the book.

GRAPH A

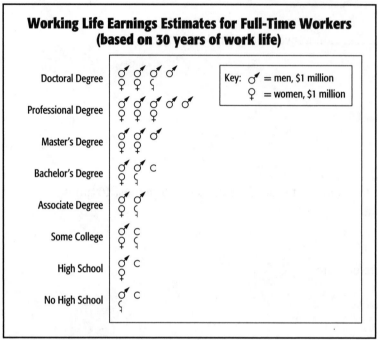

Source: U.S. Census Bureau

Answer each question by filling in the blank, answering true or false, or choosing the best multiple-choice response.

1. Graph A shows the life earnings for _____ based on education level attained.

2. From Graph A, you could infer that education helps to increase a person's chance for increased _____.

3. Men with a bachelor's degree make an average salary of over $80,000 per year.   True   False

4. Women with a high school education earn an average of $8,000 per year.   True   False

5. A female high school graduate earns about _____ than a male high school graduate over her work life.

   a. $500,000 less
   b. $500,000 more
   c. $100,000 less
   d. $1,500,000 less
   e. the same amount

6. On the average, the work life earnings of a male with an associate degree are about _____ that of a woman who has a master's degree education.

   a. $500,000 less than
   b. the same as
   c. twice as much as
   d. half of
   e. $500,000 more than

7. On the average, a woman with a master's degree will earn _____ more over her work life than a man with a high school diploma over his work life.

   a. $200,000
   b. $500,000
   c. $150,000
   d. $80,000
   e. $100,000

8. Which statement best describes Graph A?

   a. The amount of education individuals receive has little effect on their income opportunities.
   b. The amount of education individuals receive tends to affect their income opportunities.
   c. Even with an equivalent educational level, women tend to earn more money than men.

GRAPH B

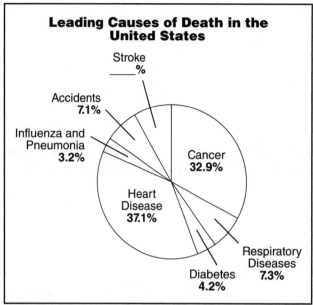

Leading Causes of Death in the United States

Stroke ____%

Accidents 7.1%

Influenza and Pneumonia 3.2%

Cancer 32.9%

Heart Disease 37.1%

Respiratory Diseases 7.3%

Diabetes 4.2%

Source: National Center for Health Statistics, 2010

Answer each question by filling in the blank, answering true or false, or choosing the best multiple-choice response.

9. Graph B represents the leading causes of _____ in the _____.

10. The percentage of deaths in the United States due to strokes is _____. (Fill this in on the graph.)

11. Heart disease and accidents are the two leading causes of death in the United States.   True   False

12. Each year heart disease accounts for slightly more than one-third of the deaths in the United States.   True   False

13. The number of deaths due to accidents is _____ less than deaths due to strokes.

    a. 1.1%
    b. 3.6%
    c. 8.2%
    d. 7.1%
    e. 13.5%

14. Together, heart disease and _____ account yearly for slightly less than three-quarters of the deaths in the United States.

    a. heart attacks
    b. strokes
    c. cancer
    d. pneumonia
    e. other

15. A diagram that correctly shows the relationship between two causes of death is:

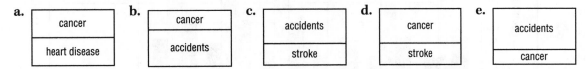

a. cancer / heart disease   **b.** cancer / accidents   **c.** accidents / stroke   **d.** cancer / stroke   **e.** accidents / cancer

16. The following statements can be concluded from Graph B EXCEPT:

    a. Cancer and heart disease cause more than 50% of the deaths in the United States.
    b. Fewer people die from accidents each year in the United States than from strokes.
    c. The leading cause of death in the United States is heart disease.
    d. Graph B shows seven leading causes of death in the United States.
    e. Research funding for heart disease exceeds the funding for all other fatal illnesses.

GRAPH C

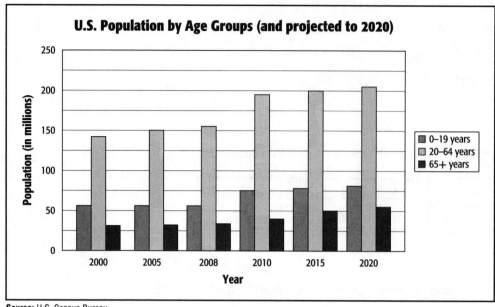

Source: U.S. Census Bureau

Answer each question by filling in the blank, answering true or false, or choosing the best multiple-choice response.

17. Graph C represents the population of the _____, and for each year shown, the population is divided into _____ age groups.

18. In 2005, the total population in the United States was approximately _____ people.

19. Graph C shows the world population over a 22-year period.   True   False

20. By the year 2005, there were about 150 million people who were 20 to 64 years old.   True   False

21. Between 2000 to 2008, approximately what was the total increase in population in the United States?

   a. 45 million
   b. 2 million
   c. 10 million
   d. 25 million
   e. 65 million

22. Between 2008 and 2010, it is shown that the _____ age group will show the greatest gain in total number of people.

    a. 0–19
    b. 20–64
    c. 65+

23. Between 2000 and 2020, it is estimated that the number of people in the 65+ age group will

    a. increase rapidly
    b. decrease rapidly
    c. consistently increase
    d. increase until 2007

24. Which conclusion can be drawn from Graph C?

    a. There is a projected steady increase in all age groups over the next ten years.
    b. Since 2010, older people are retiring at an earlier age than before 2000.
    c. The total number of people in each age group shown is expected to be greater in the year 2005 than in any previous year.
    d. There were more teenagers in 2000 than the projections show for 2020.

GRAPH D

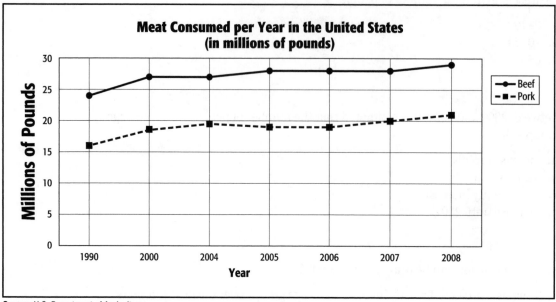

Meat Consumed per Year in the United States (in millions of pounds)

Source: U.S. Department of Agriculture

Answer each question by filling in the blank, answering true or false, or choosing the best multiple-choice response.

25. Graph D shows the amount of _____ and _____ consumed in the U.S.

26. The highest amount of beef consumed was _____ million pounds in the year _____.

27. The meat consumed in Graph D represents the number of dollars spent for every 1 million pounds of meat.   True   False

28. The amount of beef consumed in the U.S. rose steadily from the years 2005 to 2007.   True   False

29. Between what years did pork consumption in the U.S. decrease?

    a. 2007 to 2008

    b. 2006 to 2007

    c. 2004 to 2005

    d. 2000 to 2004

    e. 1990 to 2000

30. During 2008, _____ million pounds more of beef than pork were consumed.

    a. 8

    b. 1

    c. 3

    d. 7

    e. 1.5

31. Which diagram best represents information from Graph D?

    a. Beef   **b.** Pork   **c.** Beef   **d.** Pork   **e.** Beef

    1990   2004   2006   2007   2005   2006   2005   2008   1990   2008

32. Which conclusion can be drawn from Graph D?

    a. Meat prices declined during the 1990s.

    b. In the 2000s, farmers earned more income from beef than from pork.

    c. Pork and beef consumption will rise and fall with a person's spendable income.

    d. The consumption of pork can be expected to rise above the consumption of beef during the next several years.

    e. The amount of pounds of beef and pork consumed is increasing.

GRAPH REVIEW CHART

Circle the number of any problem that you missed and review the appropriate pages. A passing score is 27 correct answers. If you miss more than 5 questions, you should review this chapter.

Problem Numbers	Skill Area	Practice Pages
1, 2, 3, 4, 5, 6, 7, 8	pictograph	16–31
9, 10, 11, 12, 13, 14, 15, 16	circle graph	16–19, 32–43
17, 18, 19, 20, 21, 22, 23, 24	bar graph	16–19, 44–55
25, 26, 27, 28, 29, 30, 31, 32	line graph	16–19, 56–67

SCHEDULES AND CHARTS

Schedule and Chart Skills Inventory

This inventory allows you to measure your skills in reading and interpreting schedules and charts. Correct answers are listed by page number at the back of the book.

CHART I

International Time Table: Time Difference in Phone Dialing**				
Time Difference from U.S. Mainland to:	**U.S. Time Zones in Standard Time**			
	Eastern (New York)	**Central (Chicago)**	**Mountain (Denver)**	**Pacific (Los Angeles)**
Paris (France)	6	7	8	9
Munich (Germany)	6	7	8	9
Athens (Greece)	7	8	9	10
Peking (China)	13	14	15	16
Tel Aviv (Israel)	7	8	9	10
Tokyo (Japan)	14	15	16	17
Auckland (New Zealand)	18	19	20	21
Durban (South Africa)	7	8	9	10
Stockholm (Sweden)	6	7	8	9
London (England)	5	6	7	8
Caracas (Venezuela)	1	2	3	4
**To compute time change, add the number of hours shown to your watch.				

Answer each question below.

1. There are four United States time zones listed on Chart I: Eastern, _____, _____, and Pacific.

2. If you drive from the Eastern time zone to the Pacific time zone, you would need to change your watch by _____ hours.

3. The time difference between Chicago and Tokyo is 19 hours.   True   False

4. When it is 3 o'clock P.M. in New York, it is 8 o'clock P.M. in London.   True   False

5. When it is 12 o'clock noon Tuesday in Denver, it is 8 o'clock _____ in Auckland, New Zealand.
   a. Tuesday morning
   b. Tuesday evening
   c. Wednesday evening
   d. Wednesday morning

SCHEDULE I

Parcel Post Rate Schedule (for packages over 1 pound)

Weight not over (pounds)	Zones 1 and 2 0–150 Miles	Zone 3 151–300 Miles	Zone 4 301–600 Miles	Zone 5 601–1000 Miles	Zone 6 1001–1400 Miles	Zone 7 1401–1800 Miles	Zone 8 1801+ Miles
	ZONES BASED ON MILES PARCEL IS SHIPPED						
1	$4.90	$4.90	4.90	4.90	4.90	4.90	4.90
2	$4.90	5.15	5.70	7.02	7.33	7.62	8.09
3	$5.45	6.20	7.05	8.18	8.85	9.27	9.92
4	$6.05	7.05	7.94	9.20	10.05	10.66	11.57
5	$6.80	8.10	9.02	10.05	10.78	11.37	12.34
6	$7.50	9.20	9.94	10.94	11.45	12.02	13.03
7	$8.25	10.00	10.49	11.80	12.36	13.01	14.15
8	$8.56	10.34	10.86	12.17	12.96	13.74	15.03
9	$8.87	10.69	11.24	12.55	13.56	14.47	15.90
10	$9.18	11.03	11.61	12.92	14.16	15.21	16.77
11	$9.49	11.38	11.99	13.29	14.76	15.94	17.65
12	$9.80	11.72	12.36	13.66	15.36	16.67	18.52

Answer each question by filling in the blank, answering true or false, or choosing the best multiple-choice response.

6. Parcel post rates are determined by the _____ in which the parcel is shipped and the _____ of the parcel.

7. Schedule I gives the cost of mailing a package that weighs no more than _____ pounds.
   (number)

8. According to Schedule I, mailing a 7-pound 4-ounce package to a city 850 miles away costs $12.17.   True   False

9. The postage for mailing a package to a city 350 miles away is listed under Zone 5.   True   False

10. The postage to mail an 8-pound package 1,105 miles is _____ more than a 12-pound package 225 miles away.

    a. 99¢

    b. 93¢

    c. $1.54

    d. $1.24

    e. 89¢

11. The cost of mailing two 5-pound packages 525 miles is _____ more than mailing one 10-pound package the same distance.

    a. $3.64

    b. $1.57

    c. $5.20

    d. $6.43

    e. $3.25

SCHEDULE AND CHART SKILLS INVENTORY CHART

Use this inventory to see what you already know about schedules and charts and what you need to work on. A passing score is 9 correct answers. Even if you have a passing score, circle the number of any problem that you miss and turn to the practice pages indicated for further instruction.

Problem Numbers	Skill Area	Practice Pages
1, 2, 3, 4, 5	chart	80–87, 92–93
6, 7, 8, 9, 10, 11	schedule	80–85, 88–92, 94–95

What Are Schedules and Charts?

Schedules and charts show specific values by listing numbers and words in columns and rows.

Comparing Schedules and Charts

Look at the examples of a schedule and a chart below. Schedules and charts may look very much alike. However, they may have different uses.

EXAMPLE   Bus Schedule

Bus Schedule &

#22 LCC EXPRESS

LEAVE 10th & Willamette	19th & Pearl	30th & Hilyard	ARRIVE Lane Commnty College
7:05	7:09	7:13	7:21
7:35	7:39	7:43	7:51
8:05	8:09	8:13	8:21
8:35	8:39	8:43	8:51
9:05	9:09	9:13	9:21
9:35	9:39	9:43	9:51
10:05	10:09	10:13	10:21
10:35	10:39	10:43	10:51
11:05	11:09	11:13	11:21
11:35	11:39	11:43	11:51
12:05	12:09	12:13	12:21
12:35	12:39	12:43	12:51
1:05	1:09	1:13	1:21
1:35	1:39	1:43	1:51
2:05	2:09	2:13	2:21
2:35	2:39	2:43	2:51
3:05	3:09	3:13	3:21
3:35	3:39	3:43	3:51
4:05	4:09	4:13	4:21
4:35	4:39	4:43	4:51

Schedules often list times of events. Therefore, a useful definition of a schedule is

> a list of facts and relations primarily dealing with *times* of events.

Examples of other schedules:

 train
 television
 sports events
 school registration

EXAMPLE   Nutrition Chart

NUTRITION CHART OF FOOD GROUPS FOR 3–OUNCE SERVINGS		
Food	**Calories**	**Protein (g)**
Chicken Breast	140	27
Lamb Chop	235	22
Beef Steak	240	23
Ham Steak	205	18
Pork Chop	250	21

Source: U.S. Department of Agriculture

Charts are used to compare values of one item with another. Therefore, a useful definition of a chart is

> a list of information on which *items* are compared.

Examples of other charts:

 calorie
 weight
 mileage
 medical costs

Parts of Schedules and Charts

Schedules and Charts Have Titles

The **title** is a short description of the topic or main idea.

Bus Schedule ♿
#22 LCC EXPRESS

**NUTRITION CHART OF FOOD GROUPS
FOR 3–OUNCE SERVINGS**

Schedules and Charts Often Contain Tables

A **table** is a list of words and numbers written in rows and columns.
Columns are read up and down. Rows are read across. (You will learn how to
read tables on pages 84 and 85.)

LEAVE 10th & Willamette	19th & Pearl	30th & Hilyard	ARRIVE Lane Commnty College
7:05	7:09	7:13	7:21
7:35	7:39	7:43	7:51
8:05	8:09	8:13	8:21
8:35	8:39	8:43	8:51
9:05	9:09	9:13	9:21
9:35	9:39	9:43	9:51
10:05	10:09	10:13	10:21
10:35	10:39	10:43	10:51
11:05	11:09	11:13	11:21
11:35	11:39	11:43	11:51
12:05	12:09	12:13	12:21
12:35	12:39	12:43	12:51
1:05	1:09	1:13	1:21
1:35	1:39	1:43	1:51
2:05	2:09	2:13	2:21
2:35	2:39	2:43	2:51
3:05	3:09	3:13	3:21
3:35	3:39	3:43	3:51
4:05	4:09	4:13	4:21
4:35	4:39	4:43	4:51

Food	Calories	Protein (g)
Chicken Breast	140	27
Lamb Chop	235	22
Beef Steak	240	23
Ham Steak	205	18
Pork Chop	250	21

Source: U.S. Department of Agriculture

This table compares nutritional values of
five food groups.

This table gives the times when the bus leaves
and arrives at specific locations.

Schedules and Charts Often Have Symbols

Symbols provide additional information. A **key** is often used to give the meaning
or value of a symbol.

♿ This symbol means that the bus has
a lift to pick up riders in wheelchairs.

(g) This symbol stands for the word *grams*—the
weight unit that is used to measure protein.

Types of Questions

In this section of the book, you will be asked three types of questions to find information on charts and schedules.

Scanning the Chart (or Schedule) Questions

"Scanning the Chart (Schedule) Questions" requires looking at the main topics and information on a chart or schedule. These questions are answered by filling in words to complete the sentence. In scanning a chart (schedule),

- Pay attention to the titles, names of columns and rows, and identified units of measurement.
- Check the source of the chart's (schedule's) information.

EXAMPLE   Chart A gives the recommended daily dietary allowance of Vitamin _____.

ANSWER: **C.** Scan the titles of the columns for this information.

Reading the Chart (Schedule) Questions

"Reading the Chart (Schedule) Questions" involve locating specific information on the chart or schedule. When reading a chart (schedule), read horizontally (from left to right) and vertically (from top to bottom).

EXAMPLE   It is recommended that a boy, age 12, consume 56 grams of protein daily.

True   False

ANSWER: **False.** Starting from the left side of the chart and at the row title Males 11–14, read across to 40 in the protein column.

Comprehension Questions

Comprehension questions require you to compare values, make inferences, and draw conclusions.

CHART A

RECOMMENDED DAILY DIETARY ALLOWANCES

	Age (years)	Weight (lb)	Height (in.)	Protein (g)	Vitamin C (mg)
Infants	0.0–0.5	13	24	kg × 2.2	15
	0.5–1.0	20	28	kg × 2.0	15
Children	1–3	29	35	16	20
	4–6	44	44	25	25
	7–10	62	52	30	25
Males	11–14	99	62	40	45
	15–18	145	69	42	75
	19–22	154	70	42	75
	23–50	154	70	47	75
	51+	154	70	47	75
Females	11–14	101	62	40	45
	15–18	120	64	40	65
	19–22	120	64	40	65
	23–50	120	64	45	65
	51+	120	64	45	65
Pregnant				60	75
Lactating				65	120

Source: U.S. Food and Nutrition Board, 2010

EXAMPLE   Women 15–18 years old require _____ milligrams of Vitamin C than women 11–14 years old.

   a. 20 more   **c.** 50 more
   b. 20 less   **d.** 50 less

ANSWER: **a. 20 more.** Subtract the amount of Vitamin C for women 11–14 (45 mg) from the number for women 15–18 (65 mg).

Use Care in Reading Questions

Careful reading may help you answer "trickier" questions. The hints below will alert you to possible problem areas.

Information Is True but Not Contained on the Chart

EXAMPLE   According to Chart B, milk is more nutritional than candy.   True   False

ANSWER: **False.** While this fact is true, the chart does not compare the nutritional content of foods. Chart B only recommends the number of servings for certain food groups.

Words That Are Incorrectly Used as Possible Answers

EXAMPLE   Chart B tells

a. the percentage of certain food groups recommended for different types of people
b. the number of servings per food group recommended on a daily basis
c. the percentage of servings per food group recommended on a daily basis

ANSWER: **b.** The chart is about numbers of servings. Nothing is said about percentages, so choices **a** and **c** can be eliminated.

Symbols, Equivalents, and Abbreviations Are Often Used to Get the Correct Answer

EXAMPLE   According to Chart B, the number of servings of meat per day on a Step I diet is _____ ounces.

ANSWER: **Less than or equal to 6.** In order to provide the correct answer, find the correct meaning of the symbol, ≤. This symbol is defined at the bottom of the chart. Second, careful reading requires identification of the ounces based on a Step I diet, not a Step II diet.

CHART B

Daily Food Guide		
Food Group	**Number of Servings**	**Serving Size**
Lean Meat, poultry, fish and shellfish	≤6 ounces a day on Step I diet ≤5 ounces a day on Step II diet (leanest cuts only)	
Skim/low fat dairy foods	2–3	1 cup skim or 1 percent milk 1 cup nonfat or low fat yogurt 1 ounce low fat or fat free cheese that has 3 grams of fat or less in a serving
Eggs	≤4 yolks a week on Step I diet* ≤2 yolks a week on Step II diet*	
Fats and oils	≤6–8*	1 teaspoon soft margarine or vegetable oil 1 tablespoon salad dressing 1 ounce nuts
Fruits	2–4	1 piece fruit ½ cup diced fruit ¾ cup fruit juice
Vegetables	3–5	1 cup leafy or raw ½ cup cooked ¾ cup juice
Breads, cereals, pasta, rice, dry peas and beans, grains, and potatoes	6–11	1 slice bread ½ bun, bagel, muffin 1 ounce dry cereal ½ cup cooked cereal, dry peas or beans, potatoes, or rice or other grains ½ cup tofu
Sweets and snacks	Now-and-then	
*Includes food preparation; for fats and oils also includes salad dressings and nuts. ≤ means "less than or equal to"		

Source: American Heart Association

Schedules and Charts

Finding Information on a Table

To answer questions about schedules and charts, you will need to know how to read the columns and rows on the table. Look at the chart below.

NUTRITION CHART OF FOOD GROUPS FOR 3–OUNCE SERVINGS		
Food	**Calories**	**Protein (g)**
Chicken Breast	140	27
Lamb Chop	235	22
Beef Steak	240	23
Ham Steak	205	18
Pork Chop	250	21

Source: U.S. Department of Agriculture

Follow the steps below to answer questions about charts and schedules.

EXAMPLE   How many calories are contained in a 3-ounce serving of ham?

STEP 1   The table lists both the calories and protein found in the five types of foods. The title states that these figures are given for 3-ounce servings.

Food	Calories	Protein (g)
Chicken Breast	140	27
Lamb Chop	235	22
Beef Steak	240	23
Ham Steak	205	18
Pork Chop	250	21

STEP 2   Find the row labeled "Ham." Place a piece of paper under that row extending across the chart or use your finger as shown.

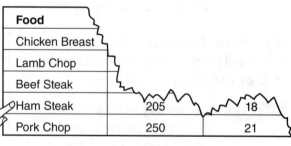

Food		
Chicken Breast		
Lamb Chop		
Beef Steak		
Ham Steak	205	18
Pork Chop	250	21

STEP 3 In the column headings, locate the title Calories. Move your eye (or finger) down that column until it meets the row labeled Ham.

Food	Calories	Protein (g)
Chicken Breast	140	
Lamb Chop	235	
Beef Steak	240	
Ham Steak	205	
Pork Chop	250	

ANSWER: There are **205 calories** in a 3-ounce serving of ham.

Answer the questions below. Check your answers to see that you are reading the chart correctly. Remember to carefully follow the steps shown above.

EXAMPLE 1 There are twenty-two grams of protein in a 3-ounce serving of

_____.

ANSWER: **lamb.** In this case, you locate the information within the table and find your answer as the row title. Find 22 in the protein column and then look across to the row labeled Lamb Chop.

EXAMPLE 2 A 6-ounce serving of pork chops would contain 250 calories.   True   False

ANSWER: **False.** You must multiply the given number of calories in a 3-ounce serving (250) by 2 to get the total number of calories in a 6-ounce serving. (250 × 2 = 500)

Practice Chart I

Chart I is a table that contains information about the importance of serving size and amount of calories necessary to maintain a balanced diet and good health. Often charts have two or more sections from which information can be drawn.

CHART I

Serving Sizes	
To make the most of maintaining a balanced diet, you need to know what counts as a serving.	
Food Group	**Serving Size**
Bread, Cereal, Rice and Pasta	1 slice bread, 1 ounce ready-to-eat cereal, $\frac{1}{2}$ cup cooked cereal, rice or pasta, 5–6 small crackers
Vegetable	1 cup raw, leafy vegetables, $\frac{1}{2}$ cup cooked or chopped raw vegetables, $\frac{3}{4}$ cup vegetable juice
Fruit	1 medium apple, banana or orange, $\frac{1}{2}$ cup chopped, cooked, or canned fruit, $\frac{3}{4}$ cup fruit Juice
Milk, Yogurt and Cheese	1 cup milk or yogurt, $1\frac{1}{2}$ ounces natural cheese, 2 ounces processed cheese
Meat, Poultry, Fish, Dry Beans, Eggs and Nuts	1–3 ounces cooked lean meat, poultry or fish. Foods which count as 1 ounce of meat: $\frac{1}{2}$ cup cooked dry beans, 1 egg, 2 tablespoons peanut butter, $\frac{1}{3}$ cup nuts

Sample Food Plan Per Day: Calories		
1,600 **For Many Sedentary Women and Some Older Adults**	**2,200** **Most Children, Teenage Girls, Active Women and Many Sedentary Men**	**2,800** **Teenage Boys, Many Active Men and Some Very Active Women**
Servings	Servings	Servings
Bread Group 6	9	11
Fruit Group 2	3	4
Vegetable Group 3	4	5
Milk Group 2–3	2–3	2–3
Meat Group 5 ounces	6 ounces	7 ounces

Source: Dietary Guidelines for Americans, 2000

Scanning Chart I

Fill in each blank as indicated.

1. The food groups that are listed on Chart I are bread, fruit, _____,
   _____, and _____.

2. Chart I compares the daily recommended _____ for three groups of
   children and adults.

3. The top section of Chart I identifies the serving size for each of the five
   _____ groups.

Reading Chart I

Decide whether each sentence is true or false and circle your answer.

4. Active teenage girls require about 1,600 calories per day.   True   False

5. An active teenage boy in high school is expected to consume about   True   False
   1,200 calories more per day than his retired grandmother.

6. To consume 2,200 calories per day, one might consume at least 2 cups of   True   False
   milk, an 8-ounce steak, 1 apple, 1 cup of orange juice, 1 banana, a green
   salad for lunch and dinner (equivalent to 4 cups), 1 cup of cooked cereal,
   4 slices of bread, and 20 crackers for snacks.

Comprehension Questions

Answer each question by choosing the best multiple-choice response.

7. Two servings of the milk group would consist of

   a. 3 slices of bread and 2 ounces of lean meat
   b. 1 cup of yogurt and 1 cup of milk
   c. 2 ounces of cheese and a carrot
   d. 1 cup of milk and $\frac{1}{2}$ cup of fruit

8. The amount of servings of bread for teenage boys is almost _____ the
   number of servings for sedentary women.

   a. same   b. $\frac{1}{2}$ c. two times   d. three times

9. All persons are recommended to consume the same amount of food servings of
   the _____ group per day.

   a. meat   b. milk   c. fruit   d. vegetable

10. From Chart I you can conclude the following:

   a. An increase in calories is needed as one gets older.
   b. Age and the activities of a person are the two best factors to
      determine the amount of calories needed per day.
   c. The more calories consumed per day, the more one must diet.

Practice Schedule I

Schedule I lists registration information for classes at a community college. On the basis of the first letter of his or her last name, a student reads down the column to find the appropriate row and then reads across the row to find the day, date, and time to sign up for classes.

SCHEDULE I

**TO NEW AND RETURNING STUDENTS
REGISTRATION INFORMATION**

Registration is Monday, Jan. 26 through Saturday, Jan. 31

PLEASE NOTE FOLLOWING SCHEDULE FOR OUR
STREAMLINED REGISTRATION PROCEDURES

If Your Last Name Begins with the Letter:	Please Register on:	During the Hours of:
S•T	Monday, January 26	9:00 A.M.–1:00 P.M.
C•D•E	Monday, January 26	1:00 P.M.–5:00 P.M.
O•P•Q•R	Tuesday, January 27	9:00 A.M.–1:00 P.M.
F•G•H	Tuesday, January 27	1:00 P.M.–5:00 P.M.
M•N	Wednesday, January 28	9:00 A.M.–1:00 P.M.
I•J•K•L	Wednesday, January 28	1:00 P.M.–5:00 P.M.
U•V•W•X•Y•Z	Thursday, January 29	9:00 A.M.–1:00 P.M.
A•B	Thursday, January 29	1:00 P.M.–5:00 P.M.

Open registration regardless of first letter of last name available daily between 5 and 6 P.M. and all day Friday, Jan. 30 and Saturday, Jan. 31 from 10 A.M. to 2 P.M.

JOHNSON COMMUNITY COLLEGE
927 WEBSTER ST.

FOR FURTHER INFORMATION PHONE

787-1984

Scanning Schedule I

Fill in each blank as indicated.

1. This schedule is useful for students enrolling in classes at _____.

2. Student registrations begin on _____ and end on _____.
                                           (day—date)   (day—date)

3. To find out more information about registration, a student can call the phone number _____.

Reading Schedule I

Decide whether each statement is true or false and circle your answer.

4. A new student, Henry Foster, can register on Tuesday, January 27, between 1:00 P.M. and 5:00 P.M.

   True   False

5. Mercedes Rodriquez can register on Thursday before 5:00 P.M.

   True   False

6. If Nate Fennerty does not register by Tuesday at 5:00 P.M., he will not be able to register during the semester.

   True   False

Comprehension Questions

Answer each question by choosing the best multiple-choice response.

7. The total number of registration hours on Monday is

   a. 4
   b. 8
   c. 9
   d. 10
   e. none of the above

8. If a student misses his or her assigned registration time, the student can register

   a. later that day at 6:30 P.M.
   b. the next week on Saturday, February 1
   c. on Friday, January 30

9. Sue Smith will have _____ hours in which to enroll on January 26.

   a. 3
   b. 4
   c. 5
   d. 8
   e. 9

10. If 160 students enroll by 1:00 P.M. on Monday, the average number of registrations per hour would be

   a. 70
   b. 40
   c. 160
   d. 10
   e. 80

Practice Schedule II

Schedule II lists times that trains arrive at and leave cities between Sacramento and Los Angeles, California.

For ease of reading, origin and destination stations are indicated with a "Dp" for "departs" and an "Ar" for "arrives." In addition, some stations have an "Ar" and a "Dp" indicator to tell the reader how long the train is scheduled to remain in that station. At stations without an "Ar" or "Dp" indicator, the train stops only long enough to permit passengers to board safely.

Notice that small picture symbols tell the services available on this train. For example, ☕ means that this train has sandwich, snack, and beverage service.

SCHEDULE II

Western Schedules				Reference Marks
Sacramento-Oakland-Los Angeles				ℝ All reserved train.
Read Down			**Read Up**	
15		Train Number	18	ℒ Sleeping car service.
Daily		Frequency of Operation	Daily	
ℝ ℒ ☐ ☒			ℝ ℒ ☐ ☒	✓ Club car service.
7:55 P	Dp	**Sacramento, CA** Ar	9:30 A	☒ Tray meal and beverage service.
8:13 P		Davis, CA	8:48 A	
8:41 P		Suisun-Fairfield, CA	8:21 A	
9:03 P		Martinez, CA	7:59 A	
9:32 P		Richmond, CA	7:28 A	☕ Sandwich, snack and beverage service.
9:50 P	Ar	**Oakland, CA** Dp	7:15 A	
10:10 P	Dp	Ar	7:00 A	
11:25 P	Ar	**San Jose, CA** Dp	5:45 P	☐ Checked baggage handled.
11:30 P	Dp	Ar	5:41 P	
12:55 A		Salinas, CA	4:12 A	
3:51 A	Ar	San Luis Obispo, CA Dp	1:28 A	A A.M.
3:59 A	Dp	*(Hearst Castle)* Ar	1:20 A	P P.M.
6:17 A		Santa Barbara, CA	10:45 P	
7:06 A		Oxnard, CA	9:53 P	Ar Arrive
8:18 A		Glendale, CA	8:43 P	Dp Depart
9:00 A	Ar	**Los Angeles, CA** Dp	8:25 P	

Scanning Schedule II

Fill in each blank as indicated.

1. The train stops at four cities between Sacramento and Oakland: Davis, _____, _____, and _____.

2. The number of the train going from Sacramento to Los Angeles is
_____, and the number of the train going from Los Angeles
to Sacramento is _____.

3. The symbol ✍ means that the train has _____.

Reading Schedule II

Decide whether each sentence is true or false and circle your answer.

4. Traveling from Sacramento, train #15 arrives in Los Angeles at   True   False
   9:00 A.M. the next day.

5. A train to Los Angeles leaves San Jose at 7:00 A.M.   True   False

6. Train #18 has sleeping car service.   True   False

Comprehension Questions

Answer each question by choosing the best multiple-choice response.

7. The total time it takes to travel from Sacramento to Oakland is slightly less than

   a. 4 hours
   b. 9 hours
   c. 7 hours
   d. 2 hours
   e. 1 hour

8. If you leave Los Angeles on Friday evening at 8:25 P.M., you will arrive in
   Sacramento at 9:30

   a. Saturday evening
   b. Sunday morning
   c. Friday evening
   d. Saturday morning
   e. Sunday evening

9. Train #18 stops at Oakland for _____ minutes before it leaves for
   Sacramento.

   a. 30
   b. 15
   c. 20
   d. 10
   e. 12

10. It takes a little over _____ for train #15 to travel from Sacramento to
    Los Angeles.

    a. 25 hours
    b. 13 hours
    c. 5 hours
    d. 8 hours
    e. 2 hours

Schedules and Charts: Applying Your Skills

A single person must file a tax return, known as Form 1040, in each year that his or her income is at least $9,350 for the year. Likewise, a married person filing a return with his or her spouse must file if their income is at least $18,707.

Many people choose to have their taxes prepared by a professional. Chart I compares the times for preparation, quoted fees, and actual fees of various tax preparers.

CHART I

Comparing Tax Preparers			
Tax Preparation Firm or Agency	Quote Fee	Actual Fee	Time From Appointment to Receiving Return
Hammer & Associates, Inc.	$40–45	$130	Same day
H & R Block	60–85	120	19 days
Sutton & Jones, Ltd., attorneys	200	200	16 days
Zeus Insurance Agency	25–30	75	4 days
Beneficial Tax Service	45–50	68	17 days
Frank T. Madison, Certified Public Accountant	125	120	14 days
Whitney & White, accountants	300–400	375	9 days
Internal Revenue Service	0	0	Same day
Conner & Matson, accountants	250–400	350	23 days
Jay A. Sullivan, attorney at law	25–50	100	7 days

Answer each question below.

1. Chart I compares quoted fees with _____ fees charged by income tax preparers.

2. From Chart I, it can be concluded that many tax preparers charge final fees that are beyond the initial quoted fees.   True   False

3. The IRS does not charge for tax preparation assistance.   True   False

4. For most firms listed, time to prepare taxes averages less than one week from appointment.   True   False

5. The two firms that provide same day service are

    a. Hammer and Zeus
    b. IRS and Sullivan
    c. Hammer and IRS
    d. Hammer and Sutton

6. According to Chart I, the difference between the quoted fee and the actual fee charged by H&R Block ranges from

    a. $29 to $49
    b. $37 to $67
    c. $35 to $60
    d. $38 to $68

7. If you had an appointment with the Zeus Insurance Agency on Friday, March 12th, you could expect your tax return by (**Note:** Days given on chart are business days.)

    a. Friday, March 12
    b. Tuesday, March 16
    c. Thursday, March 18
    d. Friday, March 19

8. Which diagram represents Chart I most accurately?

9. Which statement best describes Chart I?

    a. All tax preparation firms charge more than their quoted fees.
    b. The IRS does not charge for tax preparation assistance, and it calculates the lowest tax bill.
    c. Tax preparation firms vary widely in fees quoted and fees charged.
    d. The tax preparation firms that charge the least for their service save you the most money.

A **tax table** is a schedule that lists the tax to be paid by a wage earner. This tax is based on total income earned and filing status. Income tax rates are based on the principle of "graduated income tax." This principle means that the amount of tax to be paid increases as income increases.

SCHEDULE I

(taxable income) is—		And you are—			
At least	But less than	Single	Married filing jointly	Married filing sepa-rately	Head of a house-hold
			Your tax is—		
49,000					
49,000	49,050	8,444	6,519	8,444	7,109
49,050	49,100	8,456	6,526	8,456	7,121
49,100	49,150	8,469	6,534	8,469	7,134
49,150	49,200	8,481	6,541	8,481	7,146
49,200	49,250	8,494	6,549	8,494	7,159
49,250	49,300	8,506	6,558	8,506	7,171
49,300	49,350	8,519	6,564	8,519	7,184
49,350	49,400	8,531	6,571	8,531	7,196
49,400	49,450	8,544	6,579	8,544	7,209
49,450	49,500	8,556	6,586	8,556	7,221
49,500	49,550	8,569	6,594	8,569	7,234
49,550	49,600	8,581	6,601	8,581	7,246
49,600	49,650	8,594	6,609	8,594	7,259
49,650	49,700	8,606	6,616	8,606	7,271
49,700	49,750	8,619	6,624	8,619	7,284
49,750	49,800	8,613	6,631	8,613	7,296
49,800	49,850	8,644	6,639	8,644	7,309
49,850	49,900	8,656	6,646	8,656	7,321
49,900	49,950	8,669	6,654	8,669	7,334
49,950	50,000	8,681	6,661	8,681	7,346

Answer each question below.

1. The portion of the tax table shown applies to an income equal to or at least $_____ but less than $_____.

2. The income tax paid is based on four types of marital status:

   _____

   _____

   _____

   _____

3. The Tax Table shown can be used for a person who is single, married, or the head of a household.   True   False

4. The tax paid by a single woman on an income of $49,725 per year is $8,606.   True   False

5. A man who earns $49,850 a year and pays $7,321 in taxes should fall into the category of "Head of a household"   True   False

6. A single person earning $49,000 a year pays _____ less taxes than a single person earning $49,505.

    a. $125
    b. $194
    c. $220
    d. $180

7. With an income level of $49,815, a single person pays _____ than a married person filing jointly with a spouse.

    a. $2005 more
    b. $1572 more
    c. $2005 less
    d. $958 less

8. Members of the filing category who pay the most tax at every income level shown on Schedule I are

    a. head of household
    b. married filing jointly
    c. married filing separately
    d. none of the above

9. All of the following statements can be concluded from the Tax Table, Schedule I, EXCEPT:

    a. As income increases, total taxes paid in every filing category increase.
    b. Taxes paid each year are in part dependent on marital status.
    c. Married individuals filing separately pay the same amount of taxes as singles.
    d. As income decreases, taxes also decrease.
    e. Reducing the tax rate would benefit heads of household more than any other group.

Schedule and Chart Review

Do all the following problems. Work accurately, but do not use outside help. After completing the review, check your answers with the key at the back of the book.

CHART A

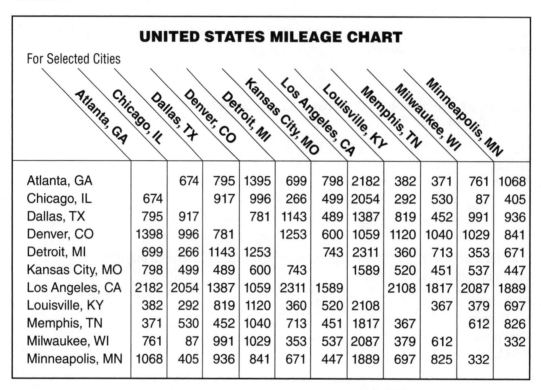

UNITED STATES MILEAGE CHART

For Selected Cities

	Atlanta, GA	Chicago, IL	Dallas, TX	Denver, CO	Detroit, MI	Kansas City, MO	Los Angeles, CA	Louisville, KY	Memphis, TN	Milwaukee, WI	Minneapolis, MN
Atlanta, GA		674	795	1395	699	798	2182	382	371	761	1068
Chicago, IL	674		917	996	266	499	2054	292	530	87	405
Dallas, TX	795	917		781	1143	489	1387	819	452	991	936
Denver, CO	1398	996	781		1253	600	1059	1120	1040	1029	841
Detroit, MI	699	266	1143	1253		743	2311	360	713	353	671
Kansas City, MO	798	499	489	600	743		1589	520	451	537	447
Los Angeles, CA	2182	2054	1387	1059	2311	1589		2108	1817	2087	1889
Louisville, KY	382	292	819	1120	360	520	2108		367	379	697
Memphis, TN	371	530	452	1040	713	451	1817	367		612	826
Milwaukee, WI	761	87	991	1029	353	537	2087	379	612		332
Minneapolis, MN	1068	405	936	841	671	447	1889	697	825	332	

Answer each question by filling in the blank, answering true or false, or choosing the best multiple-choice response.

1. The numbers on the chart represent _____.

2. Chart A is used for determining _____ between _____.

3. The mileage between Atlanta and Louisville is 382 miles.   True   False

4. There are 1,059 miles between Los Angeles and Dallas.   True   False

5. The mileage from Chicago to Denver is approximately _____ more than the mileage from Chicago to Memphis.

   a. 466 miles
   b. 212 miles
   c. 430 miles
   d. 528 miles
   e. 496 miles

6. What are the total miles driven on a trip from Dallas to Denver and then from Denver to Kansas City?

   a. 1,237 miles
   b. 1,266 miles
   c. 1,381 miles
   d. 532 miles
   e. 650 miles

7. If a trip from Los Angeles to Chicago took 5 days of driving time, approximately what is the average number of miles driven each day?

   a. 2,000 miles
   b. 300 miles
   c. 100 miles
   d. 500 miles
   e. 400 miles

8. Averaging 400 miles for each tank of gas, how many times would you need to fill the tank as you drove from Kansas City to Atlanta?

   a. 4 times
   b. 2 times
   c. 8 times
   d. 3 times
   e. 6 times

SCHEDULE A

the leisure bus ... GO GREYHOUND			
Departing Portland	**Arriving Seattle**	**Departing Los Angeles**	**Arriving San Francisco**
9:04 A.M.	1:05 P.M.	12:55 A.M.	10:30 A.M.
10:25 A.M.	2:30 P.M.	5:45 A.M.	2:40 P.M.
3:05 P.M.	7:05 P.M.	8:45 A.M.	8:05 P.M.
5:00 P.M.	9:25 P.M.	1:15 P.M.	12:55 A.M. (next day)
		3:45 P.M.	12:05 A.M. (next day)
		9:40 P.M.	8:55 A.M. (next day)

Answer each question by filling in the blank, answering true or false, or choosing the best multiple-choice response.

9. Schedule A tells only the times when buses _____ and _____ from and to major cities.

10. The bus departure schedule tells passengers the times when buses leave from the cities of _____ and _____ and arrive in the cities of _____ and _____.

11. Schedule A tells the time when buses arrive at their destinations.   True   False

12. The next bus going to Seattle after 11:45 P.M. leaves at 2:30 P.M.   True   False

13. A passenger who misses the 8:45 A.M. bus to San Francisco must wait about _____ hours for the next one.

    a. 3
    b. 6
    c. 4.5
    d. 3.5
    e. 6

14. The earliest daily bus to San Francisco leaves the station at

    a. 1:20 A.M.
    b. 6:30 A.M.
    c. 7:10 A.M.
    d. 12:55 A.M.
    e. 3:30 P.M.

15. Which graph most accurately represents Schedule A?

    a. Portland   **b.** Los Angeles   **c.** Seattle   **d.** San Francisco

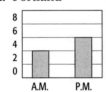

16. Seattle is located north of Portland. The same is true for San Francisco being north of Los Angeles. You can infer the following from Schedule A:

    a. While one bus goes directly to Portland, another bus with a different schedule is traveling south from Seattle.
    b. The buses that go northbound to Seattle are the same buses that are going South to Portland.
    c. Due to demand, several buses leave the station at the same time.
    d. There are more passengers going to Seattle than to Portland.
    e. There are more passengers going to San Francisco than to Seattle.

SCHEDULE B

MONTHLY PAYMENT					
Yearly Interest Schedule Rate: 5%	**Amount of Car Loan**				
No. of Months	**$10,000**	**$15,000**	**$20,000**	**$25,000**	**$30,000**
12	$856.07	$1,284.11	$1,712.15	$2,140.19	$2,568.20
24	$438.71	$658.07	$877.43	$1,096.78	$1,316.14
36	$299.71	$449.96	$599.42	$749.27	$899.13
48	$230.29	$354.44	$460.59	$575.73	$690.88
60	$188.71	$238.07	$377.42	$471.78	$566.14

Answer each question by filling in the blank, answering true or false, or choosing the best multiple-choice response.

17. Schedule B is a payment schedule for a _____ loan borrowed at _____ percent yearly interest rate.

18. The payment schedule applies to loans in the amount of $_____ to $_____.

19. A car loan of $10,000 borrowed for 12 months requires a monthly payment of $856.07.   True   False

20. To keep monthly payments below $250 on a $10,000 loan means borrowing the money for 48 months.   True   False

21. For a two-year loan, what is the difference between the monthly payments on a $15,000 loan and a $20,000 loan?

    a. $144.23
    b. $219.36
    c. $229.36
    d. $152.33
    e. $223.64

22. For a $10,000 loan, monthly payments made over two years are _____ higher than monthly payments made over four years.

    a. $55.54
    b. $178.06
    c. $208.42
    d. $261.05
    e. none of the above

23. Which graph most accurately represents Schedule B?

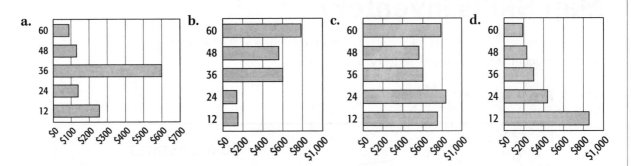

Monthly Payments on Car Loan: $10,000

24. Which statement best describes Schedule B?

  a. More loans are made for 48 months than for any other time period.

  b. Monthly payments on a car are determined by the amount of the down payment and by credit status.

  c. Monthly payments on a car loan are determined by interest rate, amount borrowed, and the amount of time for repayment.

  d. High interest rates for car loans are due to inflation.

  e. A 48-month loan costs less than a 12-month loan because the monthly payments are less.

SCHEDULE AND CHART REVIEW CHART

Circle the number of any problem that you missed and review the appropriate pages. A passing score is 20 correct answers. If you miss more than 4 questions, you should review this chapter.

Problem Numbers	Skill Area	Practice Pages
1, 2, 3, 4, 5, 6, 7, 8	chart	80–87, 92–93
9, 10, 11, 12, 13, 14, 15, 16	schedule	80–85, 88–91, 94–95
17, 18, 19, 20, 21, 22, 23, 24		

Map Skills Inventory

This inventory will help you measure your skills in reading and interpreting maps. Correct answers are listed at the back of the book.

MAP A

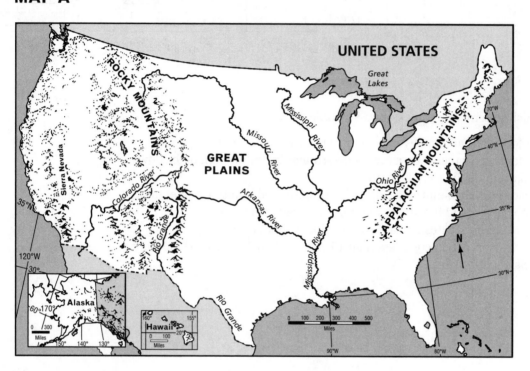

Answer each question below.

1. On Map A, the symbol ![symbol] represents _____.

2. The lines that run across the United States from left to right (labeled 30°, 35°, 40°) are called _____.

3. The Missouri River is in the northeastern United States.   True   False

4. The Rocky Mountains is a major mountain range in the western United States.   True   False

5. How many miles do the Appalachian Mountains stretch?

   a. 1,800
   b. 2,400
   c. 1,300
   d. 300
   e. 900

MAP B

Downtown Washington, D.C.

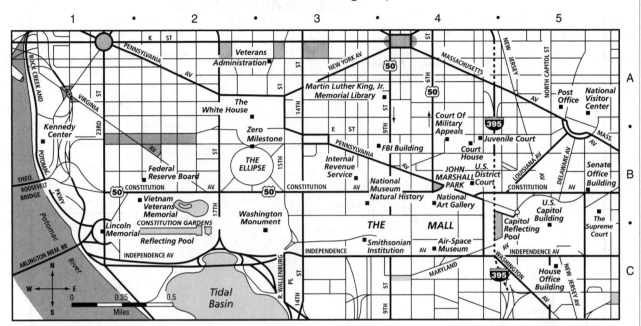

Answer each question by filling in the blank, answering true or false, or choosing the best multiple-choice response. A map-section code is given in parentheses for each place mentioned in the questions.

6. The above map shows the _____ area of Washington, D.C.

7. The Supreme Court (between B,C-5) is located about 0.25 miles _____ of the U.S. Capitol Building (C-5).
   (direction)

8. The Lincoln Memorial is located in map section C-3.   True   False

9. 17th Street crosses Highway 50 less than $\frac{1}{2}$ mile from the Washington   True   False
   Monument (C-2).

10. About how far is the distance between the U.S. Capitol Building (C-5) and the White House (A-2)?

    a. $\frac{1}{2}$ mile   **b.** 1 mile   **c.** $1\frac{1}{2}$ miles   **d.** 3 miles   **e.** 5 miles

11. Starting at the FBI Building (B-3), if you walk southeast on Pennsylvania Avenue for 0.5 mile, you end up near what building?

    a. Air-Space Museum
    b. Lincoln Memorial
    c. Internal Revenue Service
    d. U.S. District Court
    e. U.S. Capitol Building

MAP C

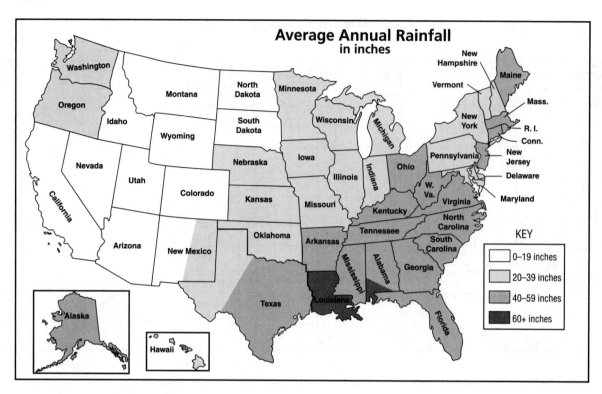

Average Annual Rainfall
in inches

KEY

	0–19 inches
	20–39 inches
	40–59 inches
	60+ inches

Answer each question by filling in the blank, answering true or false, or choosing the best multiple-choice response.

12. Map C shows the normal average _____ measured in _____ for the United States.

13. Ohio receives 40–59 inches of rain per year.   True   False

14. Only one state receives two measurably different amounts of rainfall yearly.   True   False

15. Of the following states, which state has the least average rainfall per year?

    a. Hawaii
    b. Texas
    c. Illinois
    d. Wyoming
    e. Alabama

16. Of the following states, which state shown by the map has the driest climate?

    a. Maine
    b. New York
    c. Illinois
    d. California
    e. Nebraska

MAP SKILLS INVENTORY CHART

Use this inventory to see what you already know about maps and what you need to work on. A passing score is 13 correct answers. Even if you have a passing score, circle the number of any problems that you miss and turn to the practice pages indicated for further instruction.

Problem Numbers	Skill Area	Practice Pages
1, 2, 3, 4, 5	geographical map	106–123
6, 7, 8, 9, 10, 11	directional map	106–117, 124–129
12, 13, 14, 15, 16	informational map	106–117, 130–135

What Are Maps?

A **map** is a visual display that represents the whole earth or a particular region of it. Some maps show natural features such as land masses, mountains, rivers, and oceans. Other maps show man-made features such as boundary lines between countries or states and the location of cities and highways. Maps are also used to show special information such as weather conditions and time zones.

Types of Maps

Maps are widely used in education, business, and recreation. Because of their wide variety of uses, it is convenient to group maps into three categories for study in this book: **geographical, directional,** and **informational.**

GEOGRAPHICAL MAPS

A **geographical map** shows the natural features of a region of the earth. **Natural features** are the lands, rivers, lakes, oceans, and other features that are not man-made.

The illustration below is an example of a geographical map. This map of North America shows height (elevation) by darkening or shading mountainous or other elevated land areas.

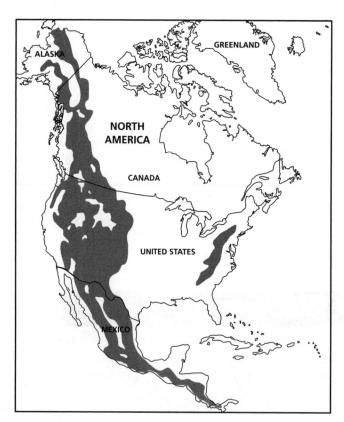

DIRECTIONAL MAPS

A **directional map** is used to show the location of cities, highways, and points of interest. Directional maps called "road maps" are commonly used by travelers to find their way across the country, across a state, or within a city. The use of these maps allows a traveler to plan a route in advance and to estimate time of travel.

The map at the right shows the main interstate highways within the state of Texas.

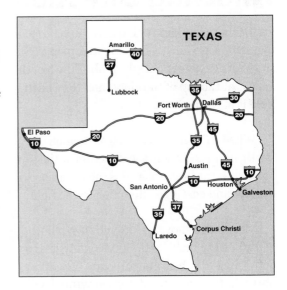

INFORMATIONAL MAPS

An **informational map** gives specific information about a particular area, or it gives information comparing different areas.

This informational map compares the birth rates in the Canadian provinces.

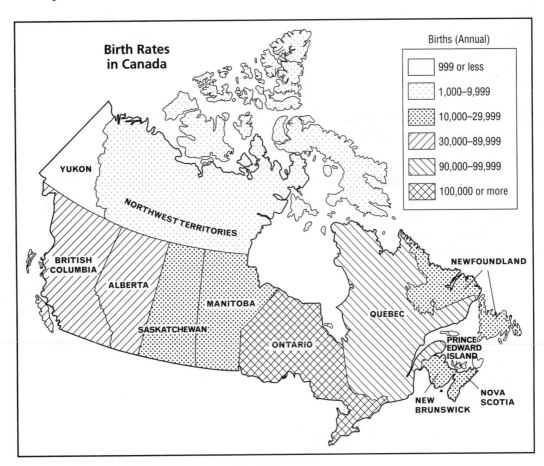

Knowing Directions on a Map

The natural shape of Earth is a sphere (ball-like shape) as shown at the right. A map in the shape of a sphere is called a **globe**. The **North Pole** is at the top of the globe, and the **South Pole** is at the bottom.

When a flat map is drawn of a region of Earth, *north* is usually the direction at the top. *South* is at the bottom, *west* is the direction to the left, and *east* is the direction to the right.

Most maps show direction by a symbol placed on the map itself. Occasionally, the top of the map may be some direction other than north. In this case, the direction symbol will indicate proper directions. Shown at the right are several common map direction symbols.

The four main map directions are often used in combination. For example, as the drawing at the right shows, point A may be both north *and* west of point O. In this case, we say that point A is *northwest* of point O. Similarly, point B is *southeast* of point O.

The example below shows several correct uses of map directions.

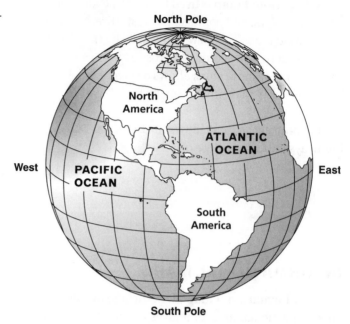

Common direction symbols

EXAMPLE   At the right is a map of Texas with locations of several cities.

- Amarillo is *northwest* of Dallas.
- Brownsville is *southwest* of Houston.
- El Paso is in *western* Texas and is *southwest* of Dallas.
- Houston is in *eastern* Texas.
- San Antonio is *southeast* of Lubbock.

As a quick way for identifying parts of the country, the United States is often divided into the following six regions: Northwest, North Central, Northeast, Southwest, South Central, and Southeast.

Become familiar with these regions before answering the questions below.

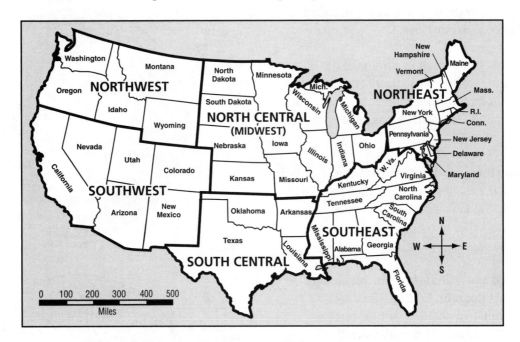

To check your understanding of map directions, fill in each blank below with one of the following directions: north, south, east, west, northeast, northwest, southeast, or southwest.

1. Nevada is _____ of New Mexico.

2. Illinois is _____ of Minnesota.

3. To find Vermont, look _____ of Indiana.

Answer the following questions about the six regions of the United States.

4. The largest region on the West Coast is the _____.

5. The region that contains the least number of states is the _____.

6. Florida is in the region known as the _____.

Understanding Longitude and Latitude Lines

To identify the location of a point on the globe, mapmakers draw two sets of lines called **longitude** and **latitude lines.**

Longitude lines are drawn from the top of the globe at the North Pole to the bottom of the globe at the South Pole. These lines are often called North-South lines. Notice that longitude lines meet at each pole.

Latitude lines are drawn across the globe from left to right. These lines are often called East-West lines. Notice that latitude lines are parallel and circle the globe; they never meet at a point and they never cross.

Longitude and latitude lines are numbered in units called **degrees.** Latitude lines are numbered from 0° to 90°. The latitude line numbered 0° is a special line called the **equator.** The equator is halfway between the North and South Poles, and it divides the globe into two parts called **hemispheres.**

The **Northern Hemisphere** consists of all places that lie north of the equator. Latitude lines are numbered from 0° at the equator to 90° at the North Pole.

The **Southern Hemisphere** consists of all places that lie south of the equator. Latitude lines are numbered from 0° at the equator to 90° at the South Pole.

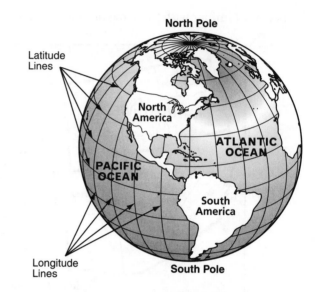

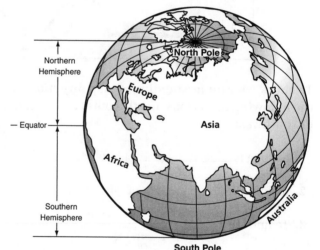

Longitude lines are numbered from 0° to 180°. The longitude line numbered 0° passes through Greenwich, England. From Greenwich, the longitude lines are numbered from 0° to 180° as you move toward the west, and from 0° to 180° as you move toward the east. The longitude line numbered 180° is exactly halfway around the earth from Greenwich.

The continents are also often said to be in the **Western Hemisphere** or in the **Eastern Hemisphere.** The globe at the top of the page shows the continents in the Western Hemisphere: North and South America. The globe in the lower half of the page shows the continents of the Eastern Hemisphere: Europe, Asia, Africa, and Australia. The continent of Antarctica is located in the Southern Hemisphere.

All the continents of Earth are shown here on a single flat map. This map is called a "Mercator Projection." Notice that the longitude lines are drawn to be parallel to each other. Although this map shows the exact locations of land and water areas of Earth, it does distort their actual sizes as you look to the far north and far south. In particular, Greenland appears to be much larger on a Mercator Projection than it is on a globe.

Become familiar with this map of Earth and then answer the questions below.

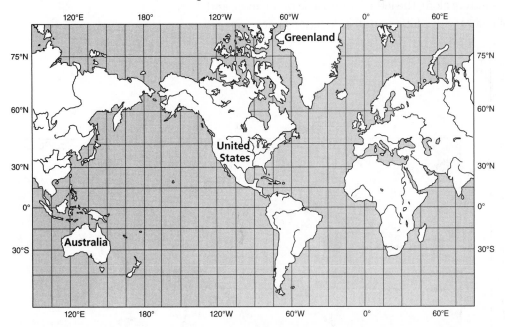

To check your understanding of longitude and latitude lines, fill in each blank below.

1. The lines that run north and south are called _____ lines.

2. The lines that run east and west are called _____ lines.

3. The latitude line that is numbered 0° is called the _____.

4. The latitude line that runs close to the southern border of the United States is the _____ line.

5. Australia lies completely _____ of the equator.
   (direction)

Using an Index Guide

On most maps, an **index guide** is used to locate countries, cities, and other points of interest. While longitude and latitude lines identify a point's location on any globe or map, an index guide is a listing that identifies a small area on a map by which a specific place is located. An index guide consists of

1. a row of numbers across the top (or bottom) of the map,
2. a column of letters along the left (or right) side of the map, and
3. a listing of places with a letter-number location code.

COUNTRIES

Albania	G-6
Austria	F-5
Belarus	D-7
Belgium	E-4
Bosnia	F-6
Bulgaria	G-7
Croatia	F-6
Czech Republic	E-6
Denmark	D-5
Estonia	C-7
Finland	B-7
France	F-3
Germany	E-5
Greece	G-7
Hungary	F-6
Ireland	D-2
Italy	G-5
Latvia	C-7
Lithuania	D-7
Macedonia	G-7
Netherlands	D-4
Norway	B-5
Poland	D-6
Portugal	G-1
Romania	F-7
Slovenia	F-5
Spain	G-2
Sweden	B-6
Switzerland	F-4
United Kingdom	D-3
Yugoslavia	G-6

MAJOR CITIES

Amsterdam (Neth.)	D-4
Athens (Gr.)	H-7
Barcelona (Sp.)	G-3
Berlin (Ger.)	D-5
Bonn (Ger.)	E-4
Brussels (Bel.)	E-4
Copenhagen(Den.)	D-5
Dublin (Ire.)	D-2
Hamburg (Ger.)	D-5
Helsinki (Fin.)	C-7
Lisbon (Por.)	G-1
Liverpool (U.K.)	D-3
London (U.K.)	D-3
Lyon (Fr.)	F-4
Madrid (Sp.)	G-2
Marseille (Fr.)	F-4
Milan (It.)	F-4
Munich (Ger.)	E-5
Oslo (Nor.)	C-5
Paris (Fr.)	E-3
Rome (It.)	G-5
Stockholm (Swe.)	C-6
Vienna (Aus.)	E-6
Zurich (Swi.)	F-4

MAJOR RIVERS

Danube	E-5, F-6
Ebro	F-2
Elbe	D-5
Loire	E-3
Po	F-4
Rhine	E-4
Rhône	F-4
Seine	E-3

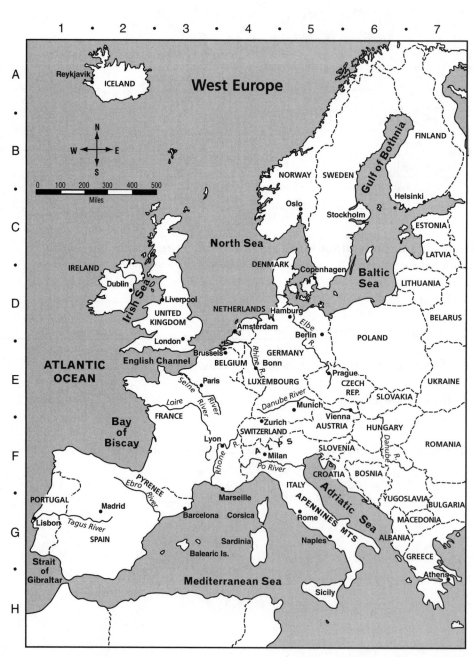

The example below shows the correct use of an index guide.

EXAMPLE　**Locate the city of Helsinki.**

STEP 1　Find the name of the city Helsinki on the index guide.

Read the letter-number code: C-7

STEP 2　Locate C on the left side of the map. Scan directly to the right, and stop below the number 7. The point of intersection of the two lines of sight (across from C and below 7) gives the area in which Helsinki is located.

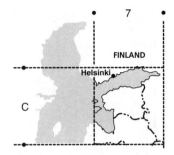

ANSWER: Helsinki is on the southern coast of Finland.

Notice that the letters and numbers are centered in the area they refer to. For instance C refers to the area halfway up to B and halfway down to D. The boundaries are marked with a dot.

Check your understanding of the use of an index guide. Use the map on the previous page to answer each question below by filling in the blank or by choosing true or false.

1. Write the letter-number code of each of the following cities:

   a. London _____

   b. Athens _____

   c. Hamburg _____

2. The Rhine River is in the country of _____.

3. The large body of water in map section C-6 is the _____.

4. Near what large city does the Elbe River run into the North Sea? _____

5. In what country is Lisbon? _____

6. The largest country that shares a border with Poland is Germany.　　　True　　　False

7. Stockholm is north of Oslo.　　　True　　　False

8. Marseille is in map section F-2.　　　True　　　False

Using a Distance Scale

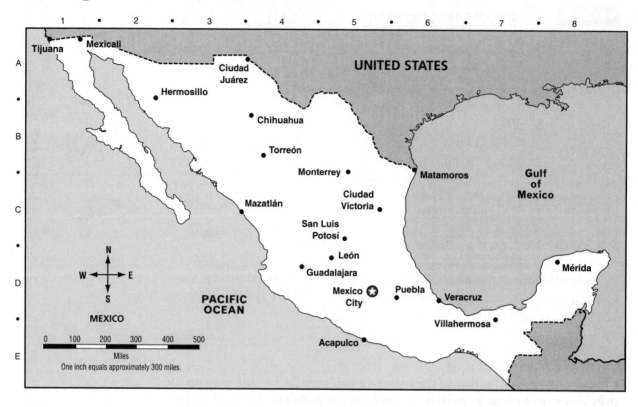

CITIES AND TOWNS

City	Grid	City	Grid	City	Grid	City	Grid
Acapulco	E-5	Hermosillo	B-2	Mexico City	D-5	Tijuana	A-1
Chihuahua	B-4	León	D-5	Monterrey	C-5	Torreón	B-4
Ciudad Juárez	A-4	Matamoros	C-6	Puebla	D-6	Veracruz	D-6
Ciudad Victoria	C-5	Mazatlán	C-3	San Luis Potosí	D-5	Villahermosa	E-7
Guadalajara	D-4	Mérida	D-8				

EXAMPLE   What is the direct distance between Torreón and Matamoros?

STEP 1   Use the index guide to find Torreón and Matamoros.

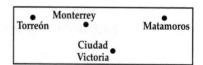

STEP 2   Use a ruler to measure the approximate distance between them: about $1\frac{1}{2}$ inches.

STEP 3   Multiply $1\frac{1}{2}$ by 300, the number of miles per inch given by the distance formula (found under the map scale).

$$300 \times 1\frac{1}{2} = 300 \times \frac{3}{2} = \textbf{450 miles}$$

ANSWER: About 450 miles

As the previous example shows, you can use a ruler to measure the **direct distance** between two points. However, the direct distance is not the **road distance.** The actual road distance between two points is always greater than the direct distance measured by a ruler. In the previous example, the actual road distance between Torreón and Matamoros is about 575 miles. Actual road mileage between two cities is often written on the map itself beside the road connecting the cities.

When you don't have a ruler available, you can use a second method to measure direct distance. You can make a **mileage ruler** by drawing a distance scale on a piece of paper. The example below shows this second method.

EXAMPLE Find the direct distance between the Mexico City and Mérida.

> **STEP 1** Use the index guide to find the two cities: Mexico City and Mérida.
>
> **STEP 2** Use the distance scale to make a mileage ruler. Then measure the direct mileage between the cities. As seen at the right, the total mileage between the cities is about **600 miles.**

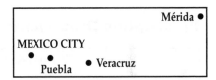

ANSWER: About 600 miles.

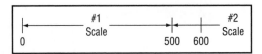

To check your understanding of the use of a distance scale, answer the question or fill in the blanks. Use a ruler or make a mileage ruler to measure direct distances on the map on page 114.

1. What is the direct mileage between each pair of cities below? Round your answer to the nearest 50 miles.

   a. Veracruz—Chihuahua _____

   b. Mazatlán—Hermosillo _____

   c. Mexico City—Villahermosa _____

   d. Guadalajara—Acapulco _____

2. Approximately how close is Mexico City to the Pacific Ocean? _____

3. Mexico City is in map section _____ and is _____ miles
   (letter-number)   (number)
   _____ of Hermosillo.
   (direction)

4. Tijuana is in map section _____ and is _____ miles
   (letter-number)   (number)
   _____ of Mazatlán.
   (direction)

Types of Questions

In this next section, you will be working with each of the three types of maps. First, though, look at the three types of questions you'll be asked in your study of maps. These questions will help you find and interpret information on each map.

Scanning the Map Questions

"Scanning the Map" questions require you to be familiar with general features concerning correct use of the map. Answer these questions by filling in words to complete a sentence. To scan a map:

- Pay attention to the title.
- Look for any map symbols, distance scales, index guides, special keys, and any other information located on or around the outside of the map.
- Be familiar with boundary lines and other special names and symbols appearing on the map itself.

Prescription Drug Thefts by Dollar Losses (in millions)

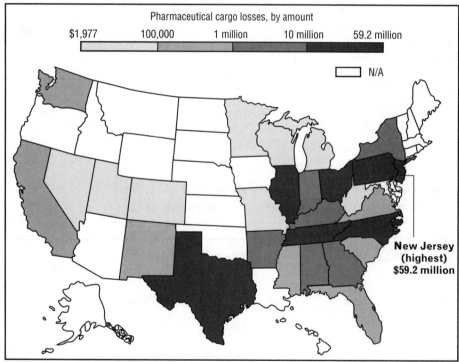

Sources: FBI; FreightWatch International

EXAMPLE 1 The above map shows the _____ _____ _____ by dollar losses across the United States over a recent four-year period.

ANSWER: prescription drug thefts (found in the title)

EXAMPLE 2 The maximum pharmaceutical cargo losses by amount is

_____

ANSWER: $59.2 million (found by looking at the listing of symbols in the key at the top of the map)

Reading the Map Questions

"Reading the Map" questions require you to locate and interpret information on the map itself. These questions are answered true or false.

EXAMPLE 1 The states with the largest drug thefts by amount are in the eastern part of the U.S.   True   False

ANSWER: True. The states with the deeper colors are found along the eastern part of the U.S. with the exception of Texas.

EXAMPLE 2 California has drug theft losses of $10 million.   True   False

ANSWER: False. California has a color that matches the key between $100,000 to $1 million, not $10 million, which would be a darker color.

Comprehension Questions

Comprehension questions require you to compare values, make inferences, and draw conclusions. Answer each question by choosing the best answer.

EXAMPLE 1 California has _____ drug theft losses in millions of dollars as the state of Washington.

  a. the same amount of
  b. fewer
  c. more
  d. four times the amount of
  e. one-half the amount of

ANSWER: a. the same amount Both states have the same color, based on the key, and thus have the same levels of drug theft losses, ranging between $100,000 to $1 million.

EXAMPLE 2 The number of states with drug theft losses between $1,977 and $100,000 is

  a. 5
  b. 7
  c. 8
  d. 10
  e. 16

ANSWER: c. 8. Count only the states marked with the lightest shading. These states are Nevada, Utah, Colorado, Minnesota, Wisconsin, Michigan, Missouri, and West Virginia.

Geographical Maps

A geographical map shows land and water formations. Often, but not always, boundary lines (borders) of countries or states and the locations of cities are also shown. Below are examples of the most commonly used geographical map symbols.

Boundary Lines

show the borders or limits of an area.

Mountains
show mountain ranges.

Rivers

show large natural streams of water.

Lakes
show bodies of water surrounded by land.

Cities •

show population centers.

State Capitals ★
show the capital of each state.

On many geographical maps, color is used to distinguish areas of different height. Land height (elevation) means "distance above sea level." **Sea level** is the level of the ocean. Thus, land that is the same level as the ocean is said to have 0 elevation, while land that is 500 feet higher than the ocean is said to have an elevation of 500 feet. The map below is an example of a geographical map.

To answer questions about geographical maps, follow the sequence below.

Scanning the Map

To scan a geographical map, identify the title, symbols, elevation key, distance scale, and longitude and latitude lines. Not all maps will have all of these items, but you should be aware of what you are working with.

EXAMPLE The map on the previous page includes the region called the
_____.

ANSWER: Southeastern United States (found in the title)

Reading the Map

To read a geographical map, find the place or area asked about. Interpret symbols and shading, and identify longitude and latitude as needed.

EXAMPLE Montgomery is the capital of Mississippi.   True   False

STEP 1   Look at the symbols for cities and decide which symbol stands for a state capital. Remember, there is only one state capital in a state.

STEP 2   Find the capital of Mississippi.

ANSWER: False. The capital of Mississippi is Jackson. Montgomery is the capital of Alabama.

Comprehension Questions

Comprehension questions require you to compare values, make inferences, or draw conclusions.

EXAMPLE What state would be the most likely choice for a mountain vacation?

a. Mississippi
b. Alabama
c. North Carolina
d. South Carolina
e. Florida

STEP 1   Look at each of the states listed above.

STEP 2   Use the map symbol for mountainous areas to decide which state contains the largest amount of mountainous terrain.

ANSWER: c. North Carolina

Practice Geographical Map

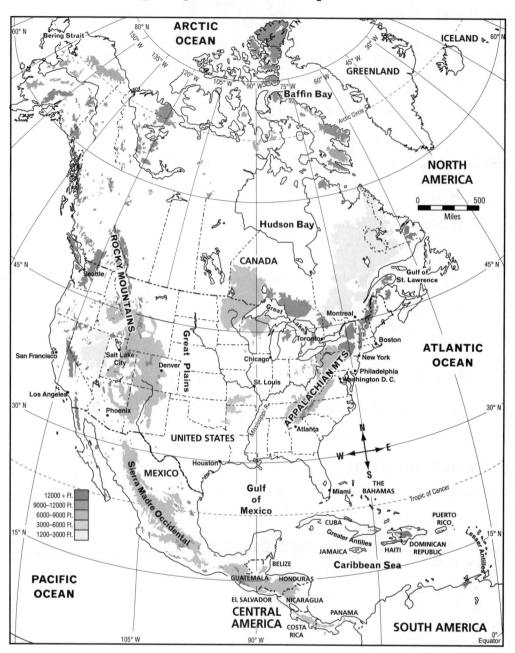

Scanning the Map

1. The dashed line (----) between 60° and 80° latitude stands for a special latitude line called the _____.

2. The map distance scale represents a total distance of _____ miles.

3. The two largest countries on the continent of North America are _____ and _____.

4. The _____ longitude line passes through the eastern edge of Cuba.

Reading the Map

Decide whether each sentence is true or false and circle your answer.

5. The United States is north of the 20° latitude line.   True   False

6. Hudson Bay is located in Canada.   True   False

7. The equator passes through the continent of North America.   True   False

Comprehension Questions

Answer each question by choosing the best multiple-choice response.

8. You can conclude from the map that the United States has _____ land area than Mexico.

   a. less
   b. less populated
   c. more
   d. more populated

9. Approximately how many miles is it from Los Angeles, California, to New York City?

   a. 1,500
   b. 2,000
   c. 2,400
   d. 3,200
   e. 5,000

10. From the map of North America, what can you conclude?

   a. North America is characterized by a uniform distribution of mountainous areas.
   b. All of the major rivers of North America eventually drain into the Atlantic Ocean.
   c. North America contains more natural resources than the continent of South America.
   d. North America is the largest of all of the continents north of the equator.
   e. North America is characterized primarily by many high mountainous areas in the West and by a few low mountain ranges in the East.

Geographical Maps: Applying Your Skills

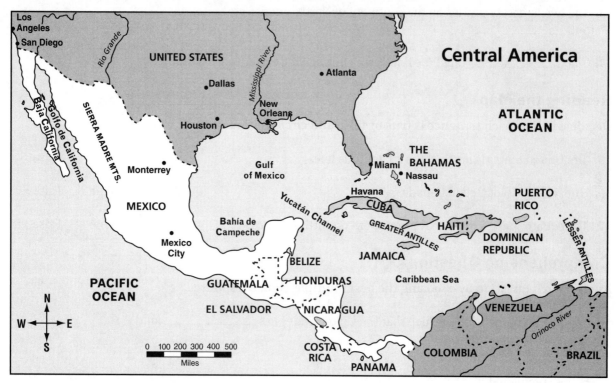

Answer each question by filling in the blank, answering true or false, or choosing the correct multiple-choice response.

1. El Salvador shares borders with the two countries: _____ and _____.

2. Mexico shares borders with two countries to the south: _____ and _____.

3. The largest island off the southern coast of Florida is _____.

4. Puerto Rico is located _____ of the Dominican Republic.
   (direction)

5. The countries of Haiti and the Dominican Republic are part of the same island.   True   False

6. Panama shares borders with Costa Rica and Colombia.   True   False

7. Havana, Cuba, is about how many miles from Miami, Florida?

    a. 50
    b. 250
    c. 450
    d. 650
    e. 850

8. Cuba is approximately how many miles from Nicaragua?

    a. 300
    b. 500
    c. 900
    d. 1,200
    e. 1,500

9. Honduras shares borders with the countries of Guatemala, _____ and _____.

    a. Belize, El Salvador
    b. Belize, Nicaragua
    c. El Salvador, Nicaragua
    d. El Salvador, Panama
    e. Panama, Nicaragua

10. To make an overland trip from Nicaragua to El Salvador requires going through what country?

    a. Belize
    b. Costa Rica
    c. Guatemala
    d. Panama
    e. Honduras

11. Looking carefully at the map, what can you conclude?

    a. Central American countries are characterized by a uniform distribution of mountainous areas.
    b. Central American countries are characterized by mountainous interior regions and narrow lowland coastal regions.
    c. Central American countries are surrounded by three large bodies of water.
    d. All of the capitals are located along the Pacific coast.
    e. The fastest travel route is from north to south.

Directional Maps

You are probably very familiar with directional maps. City maps are used to locate streets, buildings of interest, and points of interest such as parks. State and country maps are used to show the location of cities and of the highway systems that connect them. Before you work with directional maps, look at some commonly used symbols that appear on them.

Roads and highways are shown by thick or thin lines.

———— Two-lane roads

▬▬▬ Divided highway, four lanes or more

══════ Usually an interstate or U.S. highway

Cities are shown by dots and circles.

• Cities

★ State capitals

Highway route numbers are shown as numbers inside special symbols.

🛡75 Interstate highway 75

⬭59 U.S. highway 59

⑫ State highway 12

Other symbols

✕ Airport

⇒ Intersection of roads and major highways

The map below is a section of a directional map. Refer to this map as you read the example questions on the next page.

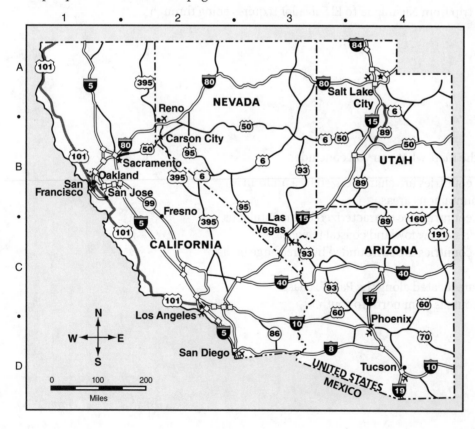

Cities	
Las Vegas	C-3
Oakland	B-1
Phoenix	D-4
Reno	B-2
Sacramento	B-1
Salt Lake City	A-4
San Diego	D-2
San Francisco	B-1
San Jose	B-1
Tucson	D-4

To answer questions about directional maps, follow the sequence below.

Scanning the Map

To scan a directional map, notice the map title, index guide, direction key, and distance scale.

EXAMPLE The map distance scale represents a total distance of
_____ miles.

ANSWER: 200 miles

Reading the Map

To read a directional map, use the index guide to locate places, and use the direction key and distance scale to identify directions and compute distances.

EXAMPLE The major interstate highway running between Phoenix and   True   False
Los Angeles is **17**.

      STEP 1   Use the index guide to find both cities.

      STEP 2   Identify the interstate highway running between them.

ANSWER: False. Interstate **10** runs between Phoenix and Los Angeles.
Notice that **17** runs north and south out of Phoenix.

Comprehension Questions

To answer comprehension questions about directional maps, you will compare locations and distances, make inferences, and draw conclusions.

EXAMPLE The direct distance between Sacramento and Los Angeles is
about _____ the direct distance between Sacramento
and Reno.

      a. 250 miles less than
      b. 100 miles less than
      c. the same as
      d. 220 miles more than
      e. 100 miles more than

      STEP 1   Measure the approximate distance between
                     Sacramento and each other city.

                     Sacramento to Los Angeles is about 350 miles.
                     Sacramento to Reno is about 130 miles.

      STEP 2   Subtract the distances.

                     $350 - 130 = 220$

ANSWER: d. 220 miles more than

Practice Directional Map

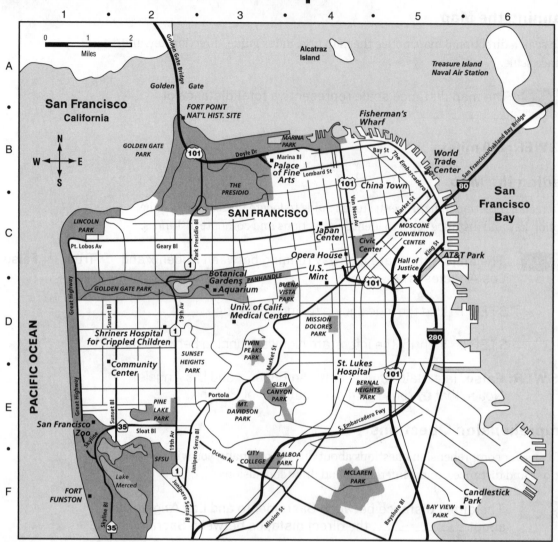

POINT OF INTEREST

Alcatraz Island	A-4
AT&T Park	C-5
Buena Vista Park	D-3
Candlestick Park	F-5
China Town	B-4
Civic Center	C-4
Community Center	D-1
Fisherman's Wharf	B-4
Glen Canyon Park	E-3
Golden Gate Bridge	A-2
Golden Gate Park	B-2
Hall of Justice	C-5
Japan Center	C-4
Lake Merced	F-2
Lincoln Park	C-1
Marina Park	B-4
Mt. Davidson Park	E-3
Opera House	C-4
Palace of Fine Arts	B-3
Pine Lake Park	E-2
San Francisco Zoo	E-1
Sunset Heights Park	D-2
U.S. Mint	C-4
World Trade Center	B-5

Scanning the Map

Fill in each blank as indicated.

1. The map shows points of interest and street names in the city of
   _____.

2. According to the index guide, Alcatraz Island is located in the map
   section _____.
   (letter-number)

3. San Francisco Bay surrounds San Francisco on both the north and
   _____ sides.
   (direction)

Reading the Map

Decide whether each sentence is true or false and circle your answer.

4. Interstate **180** is also called the Southern Embarcadero Freeway where it   True   False
   passes through the city of San Francisco.

5. Candlestick Park is more than 1 mile from San Francisco Bay.   True   False

6. The Japan Center is near Van Ness Avenue and Geary Boulevard.   True   False

7. Another name for Highway 1 is Park Presidio Blvd.   True   False

Comprehension Questions

Answer each question by choosing the best multiple-choice response.

8. The Botanical Gardens (C-2) are located in

   a. Buena Vista Park
   b. Glen Canyon Park
   c. Golden Gate Park
   d. Candlestick Park
   e. Mt. Davidson Park

9. The walking distance from the Japan Center to Fisherman's Wharf is about

   a. 1 mile
   b. 3 miles
   c. 6 miles
   d. 10 miles
   e. 20 miles

10. The directional map of San Francisco contains the following information:

    a. location and names of all streets
    b. points of interest and major highway numbers
    c. land and water elevation

Directional Maps: Applying Your Skills

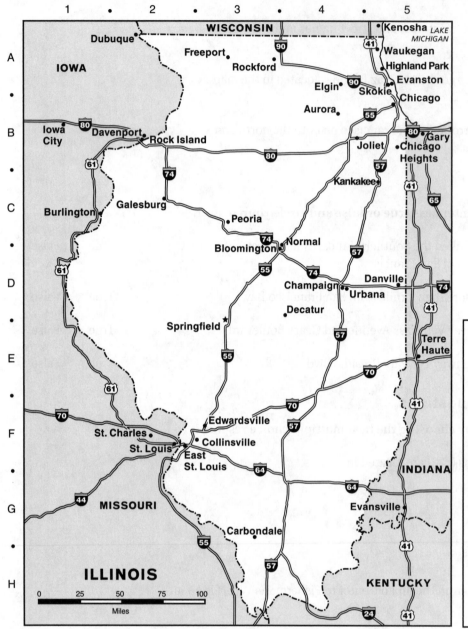

CITIES AND TOWNS

City	Grid
Aurora	B-4
Bloomington	D-3
Carbondale	G-3
Champaign	D-4
Chicago	B-5
Chicago Heights	B-5
Danville	D-5
Decatur	D-4
East St. Louis	F-2
Elgin	A-4
Evanston	A-5
Freeport	A-3
Galesburg	C-2
Highland Park	A-5
Joliet	B-4
Kankakee	C-5
Normal	D-3
Peoria	C-3
Rockford	A-3
Skokie	A-5
Springfield	E-3
Urbana	D-4
Waukegan	A-5

Answer each question by filling in the blank, answering true or false, or choosing the correct multiple-choice response.

1. According to the index guide, the city of Elgin is located in the map section

    _____.
    (letter-number)

2. The capital city of Illinois is _____.

3. The large body of water that Chicago is located near is _____.

4. Decatur is about 40 miles west of Springfield.   True   False

5. The direct distance by air between Peoria and Chicago is about 125 miles.   True   False

6. The major interstate approaching Chicago from the south is 🛡90   True   False

7. The driving distance between East St. Louis and Chicago is just a little less than

    a. 100 miles
    b. 200 miles
    c. 300 miles
    d. 400 miles
    e. 500 miles

8. If you leave Chicago driving south on 🛡57 and then turn east on 🛡74, what city will you come to?

    a. Danville
    b. Normal
    c. Peoria
    d. East St. Louis
    e. Rockford

9. Urbana is about the same distance from Chicago as it is from the city of

    a. Kankakee
    b. Decatur
    c. East St. Louis
    d. Rockford
    e. Galesburg

10. At 65 miles per hour, what is the approximate driving time between Urbana and Chicago? (**Remember:** To estimate driving time, divide the distance traveled by the average speed.)

    a. $\frac{1}{2}$ hour
    b. slightly less than1 hour
    c. under 2 hours
    d. $3\frac{1}{2}$ hours
    e. 5 hours

Informational Maps

An informational map is the most common type of map that appears in newspapers and magazines. Unlike geographical and directional maps, each informational map shows a special kind of information. Examples of informational maps include weather maps, zip code maps, time zone maps, and maps dealing with such concerns as unemployment. You will be working with several of these examples in this section.

One characteristic of informational maps is that each may have its own set of unique symbols. As an example, look at the weather forecast map below.

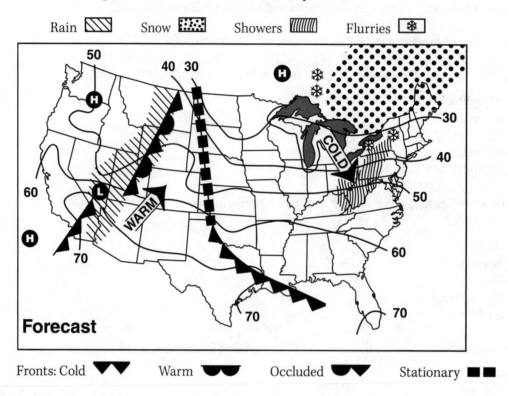

Several symbols are defined on the map, and several are defined below.

Symbols Defined on the Map

Snow ▦

Flurries: light snow and wind ❄

Rain ▨

Showers: brief periods of rain ▥

Definitions of Other Symbols

Low atmospheric pressure. A "low" indicates moist air and cool, wet weather. **L**

High atmospheric pressure. A "high" indicates dry air and warm, dry weather. **H**

(**Note:** Temperatures are indicated by numbers around the border of the U.S. The lines connecting the same temperature readings indicate locations with the same temperature. These lines are called "constant temperature lines" or **isotherms.**)

To answer questions about informational maps, follow the sequence below.

Scanning the Map

To scan an informational map, notice the graph title and any special symbols used on the map.

EXAMPLE The symbol ❄ means _____.

ANSWER: snow flurries

Reading the Map

To read an informational map, identify map symbols and other data on the map itself. Then locate the specific information needed.

EXAMPLE The high temperatures shown for most of the southeastern United States are in the 40s.   True   False

> **STEP 1** Locate the southeastern states in the lower right-hand corner of the country.
>
> **STEP 2** Read the temperatures indicated.

ANSWER: False. The high temperatures in the southeastern states are in the 50s, 60s, and, to a small extent, the 70s.

Comprehension Questions

To answer comprehension questions, compare the use of map symbols and other data on the map.

EXAMPLE This could be a weather map during the month of

> **a.** March
> **b.** May
> **c.** June
> **d.** July
> **e.** August

> **STEP 1** Notice that the moisture shown includes snow and flurries.
>
> **STEP 2** Combine this with the information that the temperatures range from the 30s to the 70s.

ANSWER: a. March

Practice Informational Map

Population Projections
by Hispanic-Origin
to 2010

New Hampshire
Maine
Vermont
Mass.
R. I.
Conn.
New Jersey
Delaware
Maryland

Washington
Montana
North Dakota
Minnesota
Wisconsin
Michigan
New York
Pennsylvania
Oregon
Idaho
South Dakota
Iowa
Ohio
W. Va
Virginia
Nevada
Wyoming
Nebraska
Illinois
Indiana
Utah
Colorado
Kansas
Missouri
Kentucky
North Carolina
California
Tennessee
South Carolina
Arizona
New Mexico
Oklahoma
Arkansas
Georgia
Texas
Louisiana
Mississippi
Alabama
Florida

Alaska

Hawaii

Population in thousands	
	0–49
	50–99
	100–499
	500–999
	1,000+

Source: US Census Bureau

Scanning the Map

Fill in each blank as indicated.

1. The map above shows the population in the year _____ for persons of _____ origin.

2. In the western part of the United States, the Hispanic population in the states of _____ and _____ are expected to reach one million in 2010.

3. The symbol ▨ stands for approximately _____ to _____ thousand persons on this map.

4. The information collected for this map was obtained from the _____.

Reading the Map

Decide whether each sentence is true or false and circle your answer.

5. The southern state with the largest Hispanic population in 2010 was Georgia.   True   False

6. In 2010, the number of Hispanic persons in Colorado was greater than Wyoming and Montana combined.   True   False

7. The number of Hispanic persons in New Mexico was greater than in the state of New York.   True   False

Comprehension Questions

Answer each question by choosing the best multiple-choice response.

8. In 2010, Utah had a larger Hispanic population than what state below?

   a. Idaho
   b. Texas
   c. North Carolina
   d. Arkansas
   e. Washington

9. Of the southern states, which state had the smallest Hispanic population in 2010?

   a. Alabama
   b. Georgia
   c. South Carolina
   d. North Carolina
   e. Florida

10. All of the following statements can be concluded from this map EXCEPT:

    a. Some of the largest Hispanic populations were in the southwestern portion of the U.S.
    b. The smallest Hispanic populations in 2010 were in the north central states.
    c. Arkansas had one of the smallest Hispanic populations in 2010.
    d. The northeastern states had the smallest Hispanic populations in 2010.
    e. None of the northwestern states had more Hispanics than California.

Informational Maps: Applying Your Skills

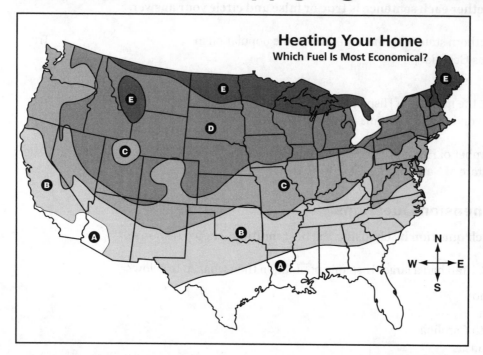

Heating Your Home
Which Fuel Is Most Economical?

Average Annual Home Heating Costs with Alternate Fuels*

*Based on a 1,200 square-foot home with average insulation and weatherization. Local climate and fuel price difference will cause heating costs to vary.

Fuel & Costs	Zone				
	A	B	C	D	E
Oil	275	687	963	1155	1375
Gas	193	484	550	754	1001
Electricity	440	1089	1529	1848	2173
Wood	180	414	624	744	876

Answer each question by filling in the blank, answering true or false, or choosing the correct multiple-choice response.

1. The table and map above compare the average annual _____ costs for each of _____ zones or areas of the country.
(number)

2. In every zone, the cost of _____ is higher than any other listed fuel.

3. The zone showing the lowest average cost for each of the fuel types is
_____.

4. Residents of Zone B who heat their home with gas can expect an average cost of $687 per year.   True   False

5. Residents of Zone B who use oil pay higher heating costs than residents of Zone C who use wood.   True   False

6. In Zone E it costs twice as much as in Zone _____ to heat with gas.

    a. A
    b. B
    c. C
    d. D

7. According to the table and map, states in Zone D pay four times as much for electricity as states in

    a. Zone A
    b. Zone B
    c. Zone C
    d. Zone D
    e. Zone E

8. The cost per year of heating with gas in Zone B is approximately _____ less than in Zone E.

    a. $300
    b. $500
    c. $700
    d. $200
    e. $100

9. In Zone D, the cost per year of heating with electricity is approximately _____ more than the cost of heating with oil.

    a. $500
    b. $600
    c. $700
    d. $900
    e. $1,000

10. Which statement best summarizes the information on this table and map?

    a. Heating costs are consistently higher in states on the West Coast than in states on the East Coast.
    b. Electricity will continue to be the most expensive source of heating costs in future years.
    c. Oil and gas used as heating fuels are more expensive in southern states than in northern states.
    d. Although costs vary from zone to zone, heating costs are consistently lower in southern states than in northern states.
    e. Many people are changing to electric heat from coal because electricity is an environmentally cleaner source of energy.

Map Review

To find out how well you understand maps, do the following problems. Answer carefully, but do not use outside help. When you've finished, check your answers with the key at the back of the book.

MAP A

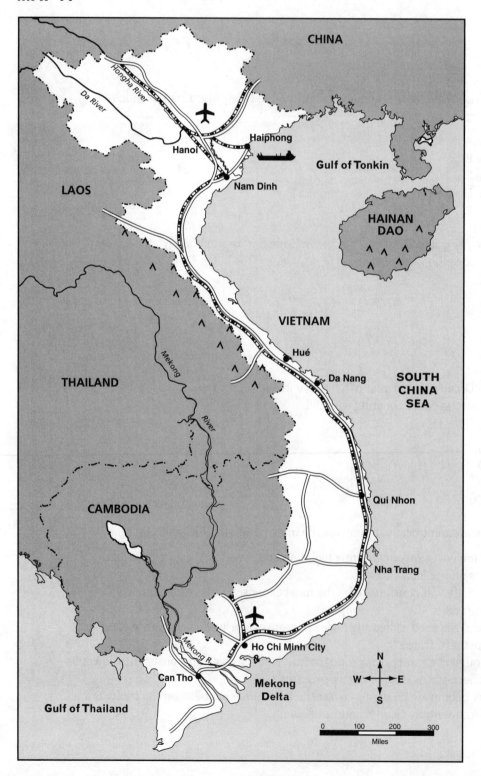

Answer each question by filling in the blank, answering true or false, or choosing the best multiple-choice answer.

1. The distance scale measures _____ miles.

2. The _____ River flows through Cambodia.

3. Ho Chi Minh City is located in southern Laos.   True   False

4. The island of Hainan Dao lies approximately 250 miles off the coast   True   False
   of Vietnam.

5. According to the map, both _____ and _____ have
   mountainous areas.

   a. Northern Laos and Cambodia
   b. Southern Laos and Hainon Dao
   c. Cambodia and Thailand
   d. Laos and Thailand
   e. Southern Laos and Cambodia

6. Laos and Vietnam are separated by

   a. the Gulf of Tonkin
   b. the Hongha River
   c. the Hanoi airport
   d. a mountain range
   e. Thailand

7. Approximately what is the distance from Can Tho to Ho Chi Minh City?

   a. 150 miles
   b. 200 miles
   c. 600 miles
   d. 3 miles
   e. 10 miles

8. The road between Haiphong and Nam Dinh

   a. is longer than the Da River
   b. crosses a mountain range
   c. is approximately the same length as the road between Hue and Da Nang
   d. crosses the Mekong River
   e. passes through Hanoi

MAP B

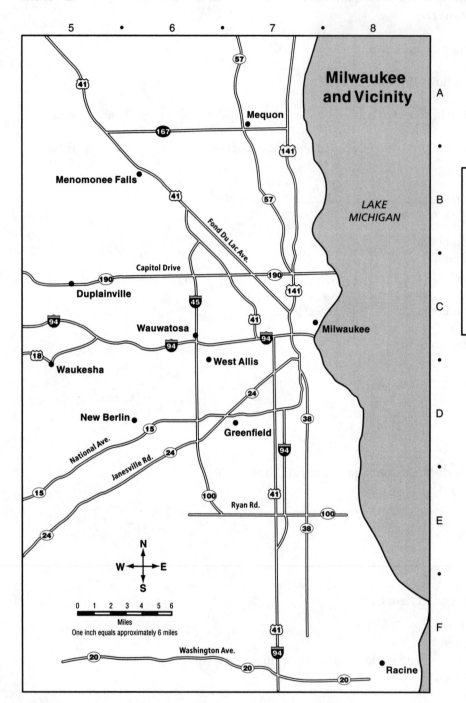

Milwaukee and Vicinity

LAKE MICHIGAN

Menomonee Falls

Mequon

Duplainville

Wauwatosa

Waukesha

West Allis

New Berlin

Greenfield

Milwaukee

Racine

Fond Du Lac Ave.

Capitol Drive

National Ave.

Janesville Rd.

Ryan Rd.

Washington Ave.

N
W   E
S

0 1 2 3 4 5 6
Miles
One inch equals approximately 6 miles

CITIES AND TOWNS	
Duplainville	C-5
Greenfield	D-7
Menomonee Falls	B-6
Mequon	A-7
Milwaukee	C-7
New Berlin	D-6
Racine	F-8
Waukesha	D-5
Wauwatosa	C-6
West Allis	D-6

Answer each question by filling in the blank, answering true or false, or choosing the best multiple-choice answer.

9. On the distance scale, 1 inch represents a distance of approximately
   _____ miles.
   (number)

10. The index guide indicates that Mequon is located in the section _____.
(letter-number)

11. Menomonee Falls is located southwest of Milwaukee.   True   False

12. Driving at 60 miles per hour, you would need to plan for about $1\frac{1}{2}$ hours   True   False
of driving time for a trip from Milwaukee to Chicago (a distance of 90 miles).

13. If you drive west from Milwaukee on 94, turn north on 45, and then turn west on
190, you will reach the town of _____.

   a. Wauwatosa
   b. Waukesha
   c. New Berlin
   d. Duplainville
   e. Greenfield

14. From this map, you can tell that Milwaukee is located near a

   a. state capital
   b. lake
   c. ocean
   d. mountain range
   e. plateau

15. Highway Route 190 is also named

   a. Fond Du Lac Avenue
   b. Ryan Road
   c. Capitol Drive
   d. National Avenue
   e. Washington Avenue

16. Traveling north on Route 41, the distance between the Route 20 turnoff and the
Route 100 turnoff is about how many miles?

   a. 5
   b. 10
   c. 50
   d. 80
   e. 100

MAP C

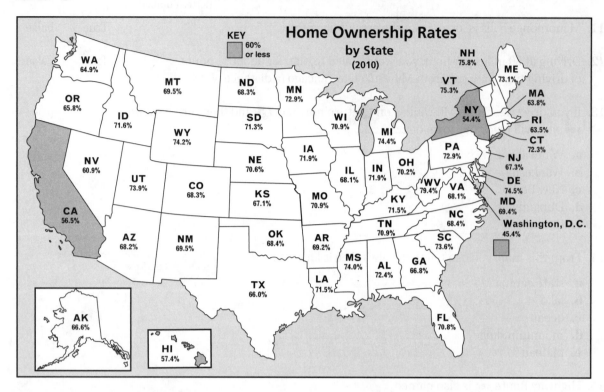

Answer each question by filling in the blank, answering true or false, or choosing the best multiple-choice response.

17. The map above shows the percent of _____ for each state in the United States during the year _____.

18. States that are shaded had home ownership rates of _____.

19. In 2010 Florida's home ownership rate was 70.8%.   True   False

20. Idaho's 2010 home ownership rate was less than 70%.   True   False

21. Which state shown had the highest home ownership rate during 2010?

 a. South Carolina
 b. Minnesota
 c. West Virginia
 d. Texas
 e. Alaska

22. In which state was the home ownership rate greater than Michigan's?

    a. Idaho
    b. Arizona
    c. Kentucky
    d. Florida
    e. Vermont

23. A family moving from Kentucky to New York is _____ to have an opportunity to own a home than a family moving from Hawaii to Mississippi.

    a. more likely
    b. less likely
    c. equally likely
    d. never likely
    e. 100% sure

24. During 2010, the home ownership rate in Alaska was _____ lower than the home ownership rate in Iowa.

    a. 4.8%
    b. 5.3%
    c. 3.5%
    d. 6.3%
    e. 6.0%

MAP REVIEW CHART

Circle the number of any problem that you missed and review the appropriate pages. A passing score is 20 correct answers. If you miss more than 4 questions, you should review the chapter.

Problem Numbers	Skill Area	Practice Pages
1, 2, 3, 4, 5, 6, 7, 8	geographical map	106–123
9, 10, 11, 12, 13, 14, 15, 16	directional map	106–117, 124–129
17, 18, 19, 20, 21, 22, 23, 24	informational map	106–117, 130–135

Posttest A

This posttest will show you how well you have learned all of the skills that you have practiced in this book. Take your time and work each problem carefully. Answer the questions by filling in the blank, answering true or false, or choosing the best multiple-choice response.

GRAPH A

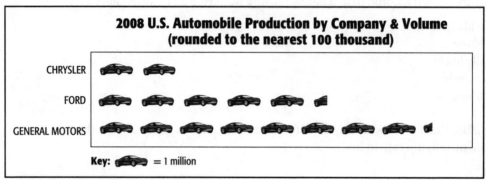

2008 U.S. Automobile Production by Company & Volume
(rounded to the nearest 100 thousand)

CHRYSLER

FORD

GENERAL MOTORS

Key: = 1 million

Source: Motor Vehicle Manufacturers Association of U.S., Inc.

1. Graph A shows the U.S. passenger car production by _____ and _____ with the symbol 🚗 representing _____ cars.

2. In 2008 General Motors Corporation sold over 10 million cars.   True   False

3. According to Graph A, for every Chrysler sold, at least _____ Fords were sold.

   a. 8
   b. 7
   c. 4
   d. 2
   e. 1

4. Which statement best describes Graph A?

   a. General Motors Corporation makes more money than any other company.
   b. Chrysler is projected to make the greatest gains in sales during the 2000s.
   c. General Motors produces twice as many cars per year as Chrysler.
   d. General Motors and Ford out-produced Chrysler by a wide margin in 2008.

GRAPH B

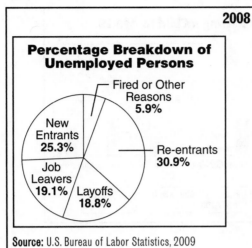

2008

Percentage Breakdown of Unemployed Persons

- New Entrants 25.3%
- Job Leavers 19.1%
- Layoffs 18.8%
- Fired or Other Reasons 5.9%
- Re-entrants 30.9%

Definitions

Re-entrants: Those who have voluntarily left their jobs and are returning once again.

Job Leavers: Those who quit their jobs and are looking for new ones.

New Entrants: Those who have never worked before, who are now looking for a job.

Source: U.S. Bureau of Labor Statistics, 2009

5. Graph B shows the percentage breakdown of _____ persons.

6. At least one-third of the unemployed are workers who are reentering the labor force, such as women who have taken time off to bear children.   True   False

7. The total percentage of unemployed who have been laid off is almost _____ those who are leaving their jobs. (**Hint:** Round percentages to the nearest ten.)

   a. one-third less than
   b. equal to
   c. three times less than
   d. double of
   e. one-half less than

8. Which statement best describes Graph B?

   a. The largest group of the unemployed is made up of new entrants.
   b. The smallest percentage of the unemployed is new entrants.
   c. The categories new entrants and re-entrants together make up the majority of the unemployed.
   d. The total percentage of those unemployed is composed of job leavers, new entrants, and re-entrants.
   e. More people are re-entering the job force than ever before.

GRAPH C

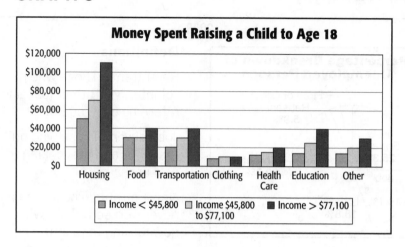

9. Graph C shows the cost of raising a child for a total of _____ years.
(number)

10. According to Graph C, the greatest increase over the years of raising   True   False
a child, regardless of family income, is in food costs.

11. Between families with incomes of $50,000 and $80,000 annually what will be the difference in the cost of educating a child until age 18?

   a. $35,000
   b. $40,000
   c. $15,000
   d. $10,000
   e. $20,000

12. All the following statements can be concluded from Graph C EXCEPT:

   a. Housing tends to be the most expensive cost in the process of raising a child to age 18.
   b. Regardless of the income of families, clothing costs are nearly the same over the 18 years that a child grows up.
   c. Housing is the most costly area across all three family income levels.
   d. The cost of raising a child will continue to increase at three times the current rate over the next decade.
   e. As income levels rise, the costs of raising a child to age 18 rises across most categories.

GRAPH D

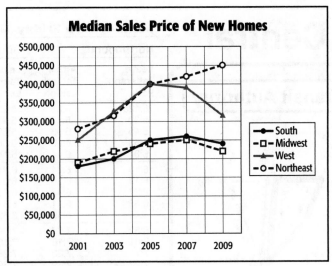

Median Sales Price of New Homes

Legend:
- South
- Midwest
- West
- Northeast

Y-axis: $0, $50,000, $100,000, $150,000, $200,000, $250,000, $300,000, $350,000, $400,000, $450,000, $500,000

X-axis: 2001, 2003, 2005, 2007, 2009

Source: U.S. Census Bureau, 2010

13. Consistently, the most expensive homes shown on Graph D are in the

_____.

14. For the years shown on the graph, 2005–2007 saw the largest decrease   True   False
in price of home sales in the West.

15. In what year did the price of homes in the West closest match the price of homes
in the Northeast?

  a. 2005
  b. 2001
  c. 2009
  d. 2003
  e. 2007

16. Which statement is true according to Graph D?

  a. By 2007 the median price of homes across the United States was
     approximately $275,000.
  b. The gap between home prices in the Northeast and the South narrowed
     considerably between 2007–2009.
  c. House prices in general are much higher in the Midwest and the South
     than in other parts of the U.S.
  d. Following price trends, the median house price in the Midwest will probably
     be as much as the median house price in the West by the year 2015.
  e. According to Graph D, a consistent decrease in median sales prices of homes
     has occurred in most regions from 2007 to 2009.

SCHEDULE A

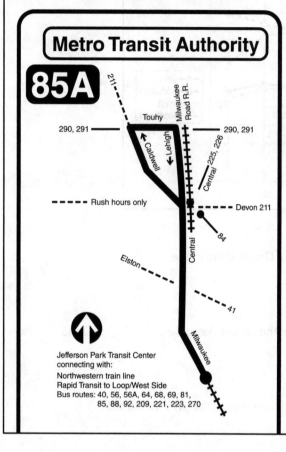

North Central

Metro Transit Authority

85A

290, 291

Touhy

Lehigh

Caldwell

Milwaukee Road R.R.

290, 291

Central 225, 226

211

- - - - Rush hours only

Devon 211

84

Central

Elston

41

Milwaukee

Jefferson Park Transit Center connecting with:

Northwestern train line
Rapid Transit to Loop/West Side
Bus routes: 40, 56, 56A, 64, 68, 69, 81,
85, 88, 92, 209, 221, 223, 270

Monday thru Friday (Morning Schedule)

Northbound		Southbound	
Leave Jefferson Park	Arrive Touhy/ Lehigh	Leave Touhy/ Lehigh	Arrive Jefferson Park
5:12a	5:25	5:25a	5:36
5:30	5:43	5:43	5:54
5:45	5:58	5:58	6:09
5:53	6:06	6:06	6:17
6:01	6:14	6:14	6:25
6:10	6:24	6:24	6:35
6:19	6:34	6:34	6:45
6:27	6:43	6:43	6:55
6:35	6:51	6:51	7:05
6:43	6:59	6:59	7:14
6:52	7:08	7:08	7:23
7:01	7:18	7:18	7:32
7:11	7:28	7:28	7:42
7:22	7:39	7:39	7:53
7:34	7:51	7:51	8:05
7:47	8:04	8:04	8:18
8:01	8:18	8:18	8:32
8:16	8:32	8:32	8:46
8:31	8:46	8:46	8:59
8:46	9:00	9:00	9:12
9:01	9:15	9:15	9:26
9:16	9:30	9:30	9:41
9:36	9:50	9:50	10:01
9:56	10:10	10:10	10:21
10:16	10:30	10:30	10:41
10:36	10:50	10:50	11:01
10:56	11:10	11:10	11:21
11:16	11:30	11:30	11:41
11:36	11:50	11:50	12:01p
11:56	12:10p	12:10p	12:21

17. Schedule A shows the bus schedule for the route number _____, which is named _____.

18. The Northbound bus leaves from Touhy/Lehigh Streets.   True   False

19. According to Schedule A, the time it takes to go from Jefferson Park to Touhy/Lehigh is

    a. 13 minutes
    b. 14 minutes
    c. 15 minutes
    d. always the same
    e. varied with the time of the morning

20. What is the total time it takes the bus leaving at 10:56 from Jefferson Park to make a round trip?

    a. 20 minutes
    b. 24 minutes
    c. 31 minutes
    d. 25 minutes
    e. 35 minutes

CHART A

A Nutritive Value of Dairy Foods						
Food Product		Calories	Protein (g)	Fat (g)	Calcium (g)	Vitamin A
Cheese-Cheddar	(1 oz)	115	7	9	204	300 (I.U.)
Cottage Cheese	(1 cup)	220	26	9	126	340
Swiss Cheese	(1 oz)	105	8	8	272	240
Whole Milk	(1 cup)	150	8	8	291	310
Nonfat Milk	(1 cup)	85	8	T	302	500
Ice Cream	(1 cup)	270	5	14	176	540
Sherbet	(1 cup)	270	2	4	103	190

Source: U.S. Department of Agriculture, *Home and Garden Bulletin No. 72*

21. Chart A is based on information from the U.S. Dept. of _____.

22. From Chart A, you can determine the number of calories in one slice of   True   False
Swiss cheese.

23. The difference in calcium between whole milk and nonfat milk is _____
grams per cup.

   a. 11
   b. 75
   c. 83
   d. 19
   e. 146

24. Which graph below most accurately reflects the chart above?

   a. calories

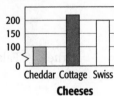

   b. protein (g)

   c. fat (g)

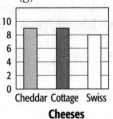

   d. calcium (g)

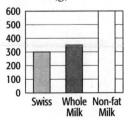

   e. vitamin A

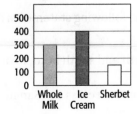

MAP A

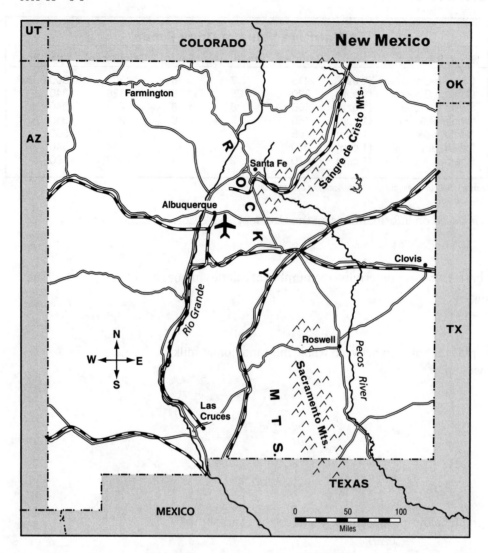

25. Map A shows the state of New Mexico, bordered by the states Utah, _____, Oklahoma, Texas, and _____.

26. There is an airport in the city of Albuquerque.   True   False

27. Where are the Sacramento Mountains located?

     a. east of the Rio Grande River   **c.** north of Albuquerque
     b. in the northeast part of the state   **d.** west of Las Cruces

28. It can be concluded from Map A that all of the following statements are true EXCEPT:

     a. The Rio Grande River crosses both the north and south borders of New Mexico.
     b. To travel from Santa Fe to Farmington, one must cross the Sangre de Cristo Mountains.
     c. The distance from Santa Fe to Albuquerque is less than 75 miles.

MAP B

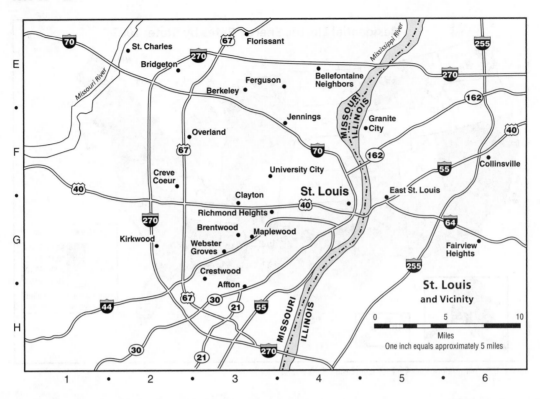

CITIES AND TOWNS		Clayton	G-3	Ferguson	E-3	Overland	F-2
Affton	H-3	Collinsville	F-6	Florissant	E-3	Richmond Heights	G-3
Bellefontaine Neighbors	E-4	Crestwood	G-3	Granite City	F-5	St. Charles	E-2
Berkeley	E-3	Creve Coeur	F-2	Jennings	F-4	St. Louis	G-4
Brentwood	G-3	East St. Louis	F-5	Kirkwood	G-2	University City	F-3
Bridgeton	E-2	Fairview Heights	G-6	Maplewood	G-3	Webster Groves	G-3

29. Webster Groves is located in map section _____.
    (letter-number)

30. Granite City is located in Illinois.   True   False

31. Richmond Heights is located

    a. 3 miles northeast of University City
    b. less than 20 miles west of St. Louis
    c. less than 20 miles east of St. Louis
    d. 6 miles south of Florissant
    e. on the border between Missouri and Illinois

32. Affton is located at the junction of ㉚ and what other highway?

    a. ④⓪
    b. ②⑦⓪
    c. ㉑
    d. ⑬
    e. ⑮

MAP C

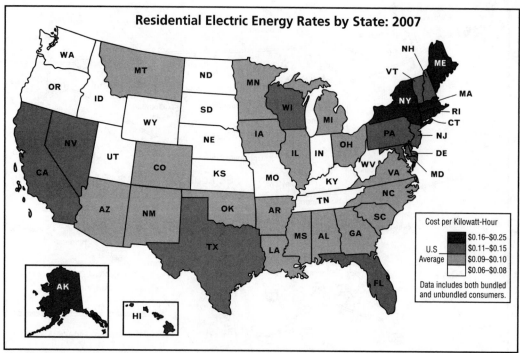

Residential Electric Energy Rates by State: 2007

Cost per Kilowatt-Hour

■	$0.16–$0.25
■	$0.11–$0.15
▨	$0.09–$0.10
□	$0.06–$0.08

U.S Average

Data includes both bundled and unbundled consumers.

Source: U.S. Energy Information Administration, Electric Sales and Revenue

33. Map C shows the _____ per kilowatt hour in states in the United States.

34. The highest electric energy costs are found in the West and the South.   True   False

35. States with the lowest electric energy costs per kilowatt hour are located in the _____ part of the United States.

   a. western
   b. northern
   c. eastern
   d. central
   e. southern

36. From Map C you can conclude the following:

   a. Families living in the northeastern states can expect to pay higher electric rates than other regions of the United States.
   b. The majority of the United States pays an average of $0.06–$0.15 per kilowatt hour for electric energy.
   c. States in the North have electric energy costs similar to that of states in the South.
   d. States with the largest populations tend to have the highest energy costs.
   e. States located in the central parts of the United States have the lowest electrical rates due to their moderate winters.

POSTTEST A CHART

Circle the number of any problem that you missed. A passing score is 30 correct answers. If you passed the test, go on to Using Number Power. If you did not pass the test, review the chapters in the book.

Problem Numbers	Skill Area	Practice Pages
1, 2, 3, 4, 5, 6, 7, 8, 9, 10, 11, 12, 13, 14, 15, 16	graphs	16–67
17, 18, 19, 20, 21, 22, 23, 24	schedules and charts	80–95
25, 26, 27, 28, 29, 30, 31, 32, 33, 34, 35, 36	maps	106–135

USING
NUMBER
POWER

Budgeting a Paycheck

Budgeting money is an important financial skill. Successful individuals and companies use budgeting to plan the flow of money.

Rather than spending money in an unplanned way, budgeting allows an individual to carefully plan the percent of income that should go for food, housing, clothing, and other important monthly costs. The circle graph below is an example of family budgeting.

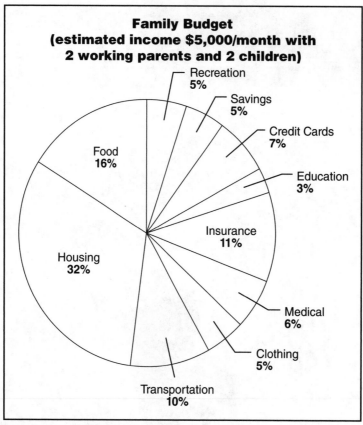

**Family Budget
(estimated income $5,000/month with
2 working parents and 2 children)**

Recreation 5%
Savings 5%
Credit Cards 7%
Education 3%
Insurance 11%
Medical 6%
Clothing 5%
Transportation 10%
Housing 32%
Food 16%

Source: Bureau of Labor Statistics

For questions 1–4, answer true or false or choose the best multiple-choice response.

1. According to the graph, the percentages for housing and food costs add up to slightly less than half of the family's total expenditures.   True   False

2. After housing and food, medical care is the largest budgeted family expenditure.   True   False

3. If a family's take-home pay was $5,000/month, the budget allows the family to spend up to _____ for food.

   a. $560
   b. $300
   c. $210
   d. $900
   e. $800

4. This family did not have any medical expenses this month. They decided to put the money budgeted for medical expenses into savings. What total percent of the family's income went into savings for the month?

   a. 21.5%
   b. 15.5%
   c. 11%
   d. 6%
   e. 5%

5. Last month, the Smith family made the following expenditures and spent over their income of $5,000/month. Using the information on the graph, circle the items that were incorrectly spent and make the necessary corrections for the following month's new budget. Remember, a "balanced" budget means total expenses equal total income.

INCOME ($5,000) — EXPENSES			
Expenses		**Adjustments (if any)**	**Corrected Budget**
Food	$800	0	$800
Clothing	$500	−$250	$250
Insurance	$550		
Housing	$1,600		
Medical	$400		
Transportation	$500		
Recreation	$250		
Savings	$250		
Credit Cards	$350		
Education	$300		
Total Expenses	$5,500		

Cost of Housing: Buying a House

Rising housing costs have made it difficult for many people to buy a home. Housing costs have been affected by a number of factors, including high interest rates for loans and increased costs of building materials.

The cost of housing varies widely across the country. It is generally recommended that no more than 25–30% of a family's income should go for housing. However, in some cities, the percentage of income that goes for housing greatly exceeds the recommended maximum percentage. As a result, many families must do without other things in order to make house payments.

	Housing Costs, Income, and Mortgage Payments			
Cities	Median Sales Price of One Family Housing*	Average Family Income per Year**	Median Monthly House Payment	Percentage of Income on House Payments
San Francisco	$493,300	$91,900	$2,450	32%
Oklahoma City	$140,500	$58,500	$1,170	24%
Denver	$219,900	$76,000	$1,267	20%
Minneapolis	$181,200	$83,900	$1,328	19%
Chicago	$199,200	$74,900	$1,373	22%
Atlanta	$123,400	$71,700	$1,075	18%
Seattle	$306,200	$80,000	$1,400	21%

*National Association of Realtors **Assuming two incomes per family

Source: U.S. Census Bureau

For questions 1–4, answer each question by filling in the blank or choosing the best multiple-choice response.

1. The median sale price of a home in San Francisco is $_____ more per year than for a family in Seattle.

2. Basing your answers on the chart, who has more money to spend after the house payment is made: a family living in Denver or a family living in Atlanta? (**Hint:** First find monthly income by dividing yearly income by the number of months per year.)

3. Based on the chart, the percentage of income paid for house payments in Minneapolis is _____ than in San Francisco.

   a. 22.0% more
   b. 15% less
   c. 31.9% more
   d. 13% less
   e. 27.5% less

4. The following conclusion can be drawn from the chart:

   a. The percentage of income spent on house payments varies regionally across the United States.
   b. Average family incomes vary according to the tax burden assumed by the family.
   c. In the seven cities listed, the percentage of income spent on housing exceeds the recommended maximum, creating a financial burden on families.
   d. The percentage of two-income families is increasing rapidly.
   e. The cost of existing housing is expected to decline.

5. Based on the chart below, decide whether you could afford the monthly house payments along with other family expenses, if you were

Living in	And Your Monthly* Income Is	And Your House Payment Is	+ Other Expenses	Answer
Oklahoma City	$3,310	$1,170	+$2,032?	Yes No
Chicago	$2,629	$1,373	+$1,820?	Yes No
Seattle	$2,482	$1,400	+$1,635?	Yes No
San Francisco	$2,622	$2,450	+$1,400?	Yes No
Minneapolis	$2,803	$1,328	+$1,470?	Yes No
*Assuming two incomes per family				

Cost of Utilities: Paying Your Electric Bill

One of the major concerns of people throughout the world is the cost of energy. Oil, gas, and electric rates have risen to record highs during the past years.

Study the information below to determine how electric rates compare from one location in the country to another.

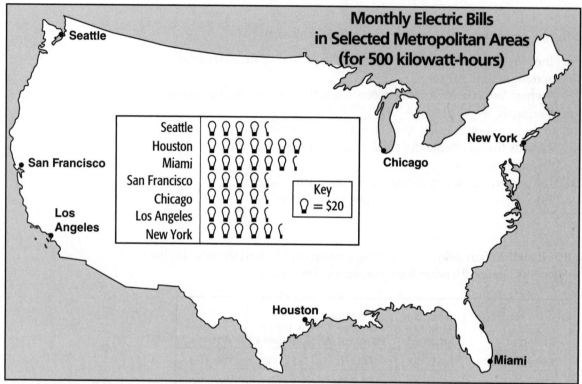

Source: U.S. Bureau of Labor Statistics

Answer each question by filling in the blank, completing the chart, or choosing the best multiple-choice response.

1. If you were living in New York City, the cost of your electric bill would be
   $_____ less per month than if you were in Houston.

2. Compute the following electric bills based on the number of kilowatt-hours. Remember the chart is based on a use of 500 kilowatt–hours.

City	Kilowatt-Hours Used	Monthly Electric Bill
Houston	750 hours	$
Seattle	1,000 hours	$
Chicago	500 hours	$
Miami	1,000 hours	$
New York City	750 hours	$

3. If electric rates increase by 10%, what would be the total amount per month for 500-kilowatt-hour electric bills in the following cities?

Seattle $_____

Miami $_____

Los Angeles $_____

4. From the graph on page 158, you could conclude the following:

a. There are more electric power sources in the West, thus accounting for cheaper costs.
b. More people use electricity in the East than in the West.
c. Southern and northern states show little difference in electric power rates.
d. Monthly electric bills tend to be lower in the West than they are in the East.
e. Electric bills tend to be higher in western cities than in eastern cities.

Cost of Utilities: Paying a Telephone Bill

One of the major costs in a budget is the telephone bill. Because of the competitiveness of telephone companies, each try to attract customers by adding special rates and features. For example, adding a second cell phone seldom costs what the first phone does. But be careful that you do not go beyond the limits of your contract in terms of number of hours phones are used, as these charges can certainly increase your monthly costs.

horizonwireless

P.O. BOX 55555
ANYTOWN, US 88888

Manage Your Account	Account Number	Date Due
	Invoice Number	0844706739

10041394 01 MB 0.362 **AUTO T4 0 3820 85704-702254 45 E NRWS2005

Mr. Bill Payer
1234 W. Elm St.
Tucson, AZ 87909–7022

Quick Bill Summary   Jan 21–Feb 20

Previous Balance (see back for details)	$152.98
Payment – Thank You	−$152.98
Balance Forward	$.00
Monthly Access Charges[1]	$117.97
Usage Charges[2]	
Voice	$.00
Data	$.00
Horizon Wireless' Surcharges[3] and Other Charges & Credits	$8.50
Taxes, Governmental Surcharges & Fees	$.00
Total Current Charges	$126.47

Total Charges Due by March 15, 2010 $126.47

CHARGES[1] PHONE LINE 1

Monthly Access Charges

AC Family Shareplan	60.00
Email & Web	29.99
250 Message Allowance	5.00
TEC Advanced Devices	7.99
	$102.98

Usage Charges[2]

Voice	.00
Data	.00
	$.00

Horizon Wireless Surcharges[3]

Fed Universal Service	1.64
Regulatory Charge	.07
Administrative Charge	.92
AZ Trans Priv on Sur on Tel	2.99
Pima Cnty Tran Priv Sur	.33
AZ State E911 Fee	.20
	$6.15

CHARGES[1] PHONE LINE 2

Monthly Access Charges

AC Family SharePlan Add'l line	9.99
250 Message Allowance	5.00
	$14.99

Usage Charges[2]

Voice	.00
Data	.00
	$.00

Horizon Wireless Surcharges[3]

Fed Universal Service	.30
Regulatory Charge	.07
Administrative Charge	.92
AZ Trans Priv on Sur on Tel	.78
Pima Cnty Tran Priv Sur Tele	.08
AZ State E911 Fee	.20
	$2.35

Use telephone "Quick Bill Summary" chart to answer the following questions.

1. You have accrued monthly access charges and they are itemized on the "Quick Bill Summary." You have contracted to use two cell phones, one for you and the other for your older teenager.

   Determine the total bill for the use of your first and major cell phone, and then determine the cost for the second cell phone used by your teenager.

   Cell Phone #1: $_____   Cell phone #2: $_____

2. You are concerned that you might be over-billed for the telephone services you have contracted. The Quick Bill Summary shows that there are two major sets of charges, namely, 1) Monthly Access Charges of $117.97 and 2) Horizon Wireless Surcharges of $8.50. In the table below, show the charges for each and determine if your bill is accurate by answering Question #3:

Monthly Access Charges	Monthly Wireless Surcharges
Phone #1: _____	Phone #1: _____
Phone #2: _____	Phone #2: _____
Total: _____	Total: _____

3. After analyzing this bill, have you been a) charged fairly or b) charged unfairly?

Cost of Food

One factor that has caused the rate of inflation to rise is the increase in food prices. Although it is less expensive to purchase food to prepare at home than eating out, the price of basic foods has fluctuated over the last several years.

Most families with young children build their meals to include the basic food groups of dairy, protein, and vegetables. Ground beef and milk products are among the most often purchased foods for budget conscious families. The cost of these items, along with tomatoes, which are frequently used in sauces and stews, is tracked in the chart below.

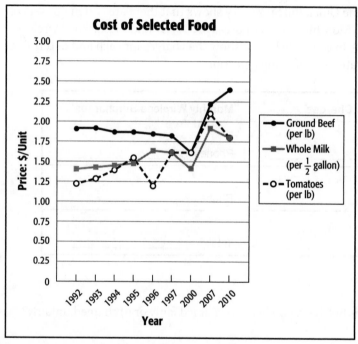

Source: U.S. Bureau of Labor Statistics, 2010

Use the graph above to answer the following questions.

1. Your family eats 5 pounds of hamburger, 2 pounds of tomatoes, and 2 gallons of milk per week. Calculate the cost of these three food items for one week at 1997 prices and at 2010 prices.

1997	2010
Beef _____	Beef _____
Milk _____	Milk _____
Tomatoes _____	Tomatoes _____
Total _____	**Total** _____

2. From question 1, find the difference in total costs for the selected food items between one week in 2010 and one in 1997.

3. From the graph, which two food items were about the same price between 1994 and 1995 and in the year 2010?

4. Compute the following food costs for 1 month if there are 4 weeks in the month and if your family consumes $4\frac{1}{2}$ pounds of beef, 2 gallons of milk, and 1 pound of tomatoes for dinner each week.

The Cost of Five Selected Foods for 1 Month: 2010

Beef	_____
Tomatoes	_____
Milk	_____
Potatoes	$5.00
Bread	$8.00
Total	_____

Costs of Operating a Car

Most people now realize that the money they pay to buy a car is only the beginning of their expenses. A buyer must also consider the cost of the daily wear and tear on the car, bills for maintaining and repairing the car, and the car's depreciation in value. These are factors in addition to monthly payments and the price of gas.

The graph below shows the relative costs of operating a car based on miles driven per year.

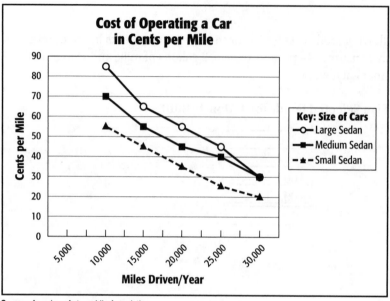

Source: American Automobile Association

Use the graph to answer the following questions.

1. The cost of operating a small car with an average of 20,000 miles driven per year plus monthly car payments of $125.00 per month would result in total expenses of _____ for the year.

2. If you drove a small car 10,000 miles during the first year, what would be the cost of operating the car? (excluding the car payments) $_____

3. Once 30,000 miles are driven in a year, the cost of operating a medium or large car becomes _____ .

    a. twice as expensive
    b. gradually increasing
    c. the same
    d. difficult to estimate
    e. none of the above

4. What would be the yearly savings in operating cost if you drove a medium sized sedan at 20,000 miles per year instead of a large sedan? $_____

5. Using the data in the graph, answer the questions related to the operating costs between the type of car and the miles driven per year.

    a. Which type of car is the most costly to operate per mile? (Circle one.)
       a small sedan with 20,000 miles ...or a large sedan with 30,000 miles

    b. How much more costly? $_____ per mile

    c. Based on your judgment and the graph, answer true or false.
       i. The more miles driven per year decreases the overall operating costs   True   False
          of a car.
       ii. The more miles driven on any type a sedan, the smaller the operating   True   False
          cost per mile.

Costs of Hospitalization

The cost of medical care is expensive and on the rise. Often the only protection against financial disaster is the purchase of a medical insurance policy. At first, monthly payments may seem an unnecessary cost, but the first major medical bill you face will seem much more manageable if you know that you will have help in paying it.

One of the main medical expenses you may face is the cost of hospitalization. The graph below compares the costs per day in a hospital to the total costs during one's stay in a hospital, which often lasts multiple days.

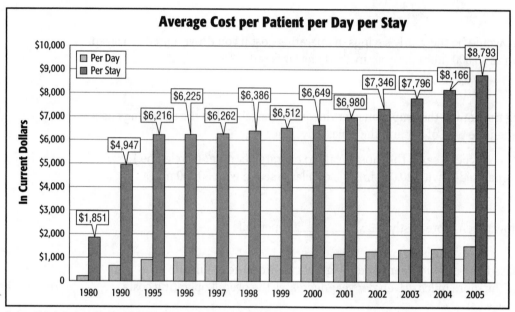

Source: U.S. Centers for Medicare & Medicaid Services, Office of the Actuary

Answer each question by filling in the blank.

1. From the graph, calculate the number of days spent in the hospital.

   a) in the year 1998 with the total bill being $6,386

   b) in the year 2005 with the total bill being $5,750

2. Suppose in 2003 you had knee surgery. The hospital you were in charged $1,250 per day. Your stay was 8 days.

What was the total expense for this hospitalization?

$_____

Your health insurance policy paid $5,617. How much did you owe after the insurance paid its portion of the bill?

$_____

3. Find the difference between the 2000 and 2005 hospital costs per stay.
   $_____ per stay

Answer true or false to the following statements, basing your answer on the graph.

4. a. Average daily costs in hospitals are growing at a faster rate than overall costs of hospital stays.   True   False

   b. No daily hospital costs increased more than $200 per stay over a one year period.   True   False

   c. Both per day costs and per stay costs showed a substantial increase between 1980 and 1990.   True   False

Calculating Driving Distances and Times

Certain maps are designed to help the traveler estimate the distance in miles as well as the approximate time necessary to reach a specific place. On the map below, the numbers indicate the mileage between two cities. For example, the distance between Los Angeles and Phoenix is 389 miles.

Often the time to drive a distance is calculated by estimating the total number of miles it takes to travel in one hour. For example, if you travel 50 miles in 1 hour, then it will take you 8 hours to travel a total of 400 miles. (50 miles per hour × 8 hours = 400 miles)

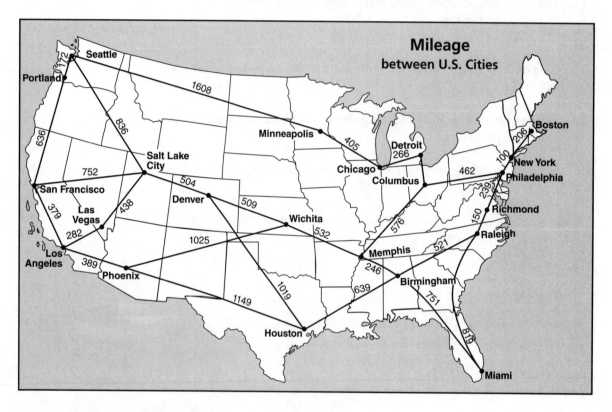

Answer each question by filling in the blank or choosing the best multiple-choice response.

1. Calculate the driving distances on the following trip. (Although many routes are possible, use the most direct route.)

From	To	Distance in Miles
Seattle	San Francisco	_____
San Francisco	Los Angeles	_____
Los Angeles	Houston	_____
	Total Miles	_____

2. Calculate the driving times for the following trips by dividing distance by miles per hour. (Round your answer to the nearest hour.)

From	To	Distance in miles	Time in hours (speed = 50 miles per hour)
Seattle	Minneapolis	_____	_____
San Francisco	Denver	_____	_____
Houston	Raleigh	_____	_____
Boston	Miami	_____	_____

3. You decide to take a trip from Phoenix to Wichita.

What is your gas mileage (in miles per gallon) if you used 50 gallons of gas (to the nearest tenth of a gallon)?

Note: $\dfrac{\text{total miles}}{\text{gallons used}} =$ miles per gallon

Answer: _____

What was your driving speed if it took 18 hours to drive this distance (round your answer to the nearest whole number)?

Note: $\dfrac{\text{total miles}}{\text{hours driven}} =$ miles per hour (speed)

Answer: _____

4. According to the map, the shortest route is from _____ to San Francisco.

a. Houston to Phoenix to Los Angeles
b. Seattle to Salt Lake City to Los Angeles
c. Wichita to Salt Lake City
d. Minneapolis to Seattle
e. Denver to Salt Lake City to Los Angeles

Reading a Directional Map

Navigating through an unknown part of a city or finding an exact address or facility in a town or city can be a challenging undertaking. Directional maps provide a tool to find a specific location on a map.

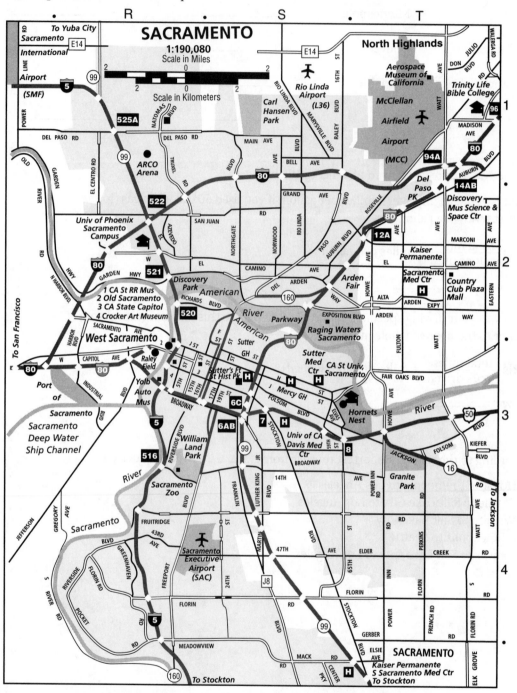

Directional maps provide certain features to assist locating specific places and destinations. In addition, they contain the following features: 1) a distance scale in miles and kilometers, 2) local, state, and interstate highways by number, 3) rivers, 4) streets, and 5) major facilities and buildings (hospitals, airports, parks, and important buildings). Numbers and letters along the horizontal and vertical sides of the map are often used in easily finding a location or destination.

Before reading the directional map, make sure you are familiar with the following diagrams and symbols:

Interstate Highway	80	River	
Highway entrance/exit		Business route	80
State routes	(16)	Airport	✈

Answer each question by filling in the blank, answering true or false, or choosing the best multiple-choice response.

1. This map shows the land area for the city of _____ in the state of _____.

2. Sacramento has two major interstate highways running through it. True False

3. The approximate distance from the Aerospace Museum (T-1) to the Sacramento Medical Center (T-2) is

    a. slightly more than one mile
    b. about 5 miles
    c. less than 4 miles
    d. 12 miles
    e. between 3 and 4 miles miles

4. From this map, use the direction to find a specific place in this city.

    a. If you drive south from West Sacramento on Interstate 5, exit and drive east on Florin Rd., and then drive north on 24th St., you will arrive at

    _____.

    b. From McClellan Airport, onto I-80 west, north on I-5/99, you will pass what popular sports center? _____

    c. From Power Inn Rd. & Elder Creek Rd. (T-4), drive north on Power, until it turns into Howe Ave., turn left on Fair Oaks Rd., and you will find what on your right? _____

5. According to the map, what is the shortest and most direct route from the Sacramento Medical Center/ Kaiser Hospital (T-2) to your home located at Stockton Blvd. and 14th Ave. (S-3)?

    a. South on Watt Ave., west on California 50, south on 65th St., west on 14th to Stockton
    b. South on Watt Ave., west on California 50, south on CA 99, east on Fruitridge, north on Stockton, to 14th
    c. South on Watt Ave, west on Florin Rd., north on 65th, west on 14th to Stockton
    d. North on Watt Ave., west on I-80, south on I-5, east on Fruitridge, north on Stockton to 14th
    e. West on El Camino Ave., south on Howe Ave., west on CA 50, south on CA 99, east on Fruitridge, north on 65th, west on 14th to Stockton

CONSTRUCTING GRAPHS

In the next pages, you will have the opportunity to analyze data and transform it into a circle, line, or bar graph, from a table or chart that is presented to you.

GRAPH A:

Task: Design a line graph with the following data.

Title: Consumption of Selected Beverages by Type: 1980 to 2007
Source: U.S. Department of Agriculture, Economic Research Service, 2010

Note: Data in gallons per person per year

	1980	1990	2000	2007
Coffee	27	27	26	30
Bottled water	3	9	17	30
Soft drinks	33	45	50	45

Design your graph here:

GRAPH B:

Task: Design a circle graph using the data given below.

Title: Sources of Sodium in U.S. Diets
Source: Institute of Medicine, Mayo Clinic, 2010

Main sources:

-Processed, prepared (fast) foods = 77%
-Natural sources = 12%
-Added while eating = 6%
-Added while cooking = 5%

Design your graph here:

GRAPH C:

Task: Design a bar graph (vertical or horizontal) using the data below.

Title: Wireless Cell Phones by Age
Source: U.S. Centers for Disease Control and Prevention, 2010

Ages	% Using Wireless Cell Phones
18–24	38%
25–29	46%
30–34	33%
35–44	22%
45–64	13%
65+	6%

Design your graph here:

GRAPH D:

Task: Design a horizontal bar graph using the data from the map below.

Title: Heart-related Deaths by States
Source: American Heart Association

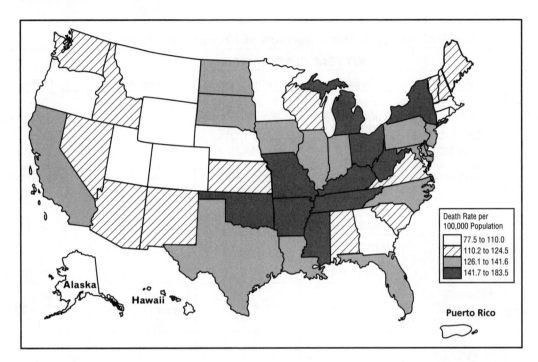

Death Rate per
100,000 Population

	77.5 to 110.0
	110.2 to 124.5
	126.1 to 141.6
	141.7 to 183.5

Alaska

Hawaii

Puerto Rico

Design your graph here:

GRAPH E:

Task: Design a bar graph (vertical) using data from the chart/table below.

Title: Number of Pharmacy Classes Offered on Days of the Week
Source: The University of Iowa, Fall 2010

The University of Iowa *College of Pharmacy Fall Semester*					
FIRST YEAR CLASS (P1) Class of 2014					
	Monday	**Tuesday**	**Wednesday**	**Thursday**	**Friday**
8:30	99:162 Biochemistry Aud 1 BSB	46:103 Fund Eval Res 100B PHAR	99:162 Biochemistry Aud 1 BSB		99:162 Biochemistry Aud 1 BSB
9:30	46:123 Pharmaceutics I 100A PHAR	69:133 Pathology (Case) 100B PHAR	46:123 Pharmaceutics I 100A PHAR	46:001 Intro Pharm Practice Reserve	46:123 Pharmaceutics I 100A PHAR
10:30		69:133 Pathology (Case) 100B PHAR	46:123 Pharmaceutics I 100A PHAR		46:50 PPL I (1) 219/226 PHAR
11:30	Communities 100 A&B PHAR	99:162 (Exam) Biochemistry Aud 1 BSB			↓ ↓ 10:30–12:20
12:30	Meetings and Speakers 100B PHAR			46:123 Pharmaceutics I 100A PHAR	46:50 PPL I (2) 219/226 PHAR
1:30	46:50 PPL I (D) 100A PHAR			46:50 PPL I (L) 100A PHAR	↓ ↓ ↓
2:30	69:133 Pathology (Lect) 1110A MERF		69:133 Pathology (Lect) 1110A MERF		—1:00–2:50—
3:30		4–5:30 P1 Exam Block 100 A&B PHAR	99:162 Biochemistry ↓	P1/P3 Exam Block 100 A&B PHAR	46:50 PPL I (3) 219/226 PHAR
4:30		Do not schedule classes or work	↓ ↓ Aud 1 BSB	Do not schedule classes or work 3:30–5:20	↓ 3:30–5:20

Design your graph here:

Posttest B

This review test has a multiple-choice format much like many standardized tests. Take your time and work each problem carefully. Circle the correct answer to each problem. When you finish, check your answers at the back of the book.

GRAPH A

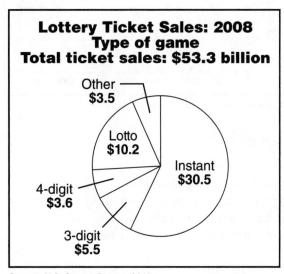

Source: U.S. Census Bureau, 2010

1. Graph A shows total lottery ticket sales by

   a. U.S. Census Bureau
   b. percent of money spent per household
   c. number of tickets sold
   d. type of game
   e. 2000 through 2008

2. Total ticket sales for 2008 was

   a. $14.2 billion
   b. $53.3 billion
   c. $21.1 billion
   d. $29.9 billion
   e. $10 billion

3. Which type of lottery game accounts for almost $\frac{2}{3}$ of total sales?

   a. 3-digit   **b.** 4-digit   **c.** Instant   **d.** Lotto   **e.** other

4. Which statement best describes the contents of Graph A?

   a. Instant lottery tickets in 1997 cost $14.20.
   b. $3.5 billion was spent on things other than lottery tickets in 1997.
   c. Lottery ticket sales in 2008 totaled $53.3 billion.
   d. Lottery ticket sales increases every year.
   e. Lotto and 4-digit lottery games were more popular than any other games.

GRAPH B

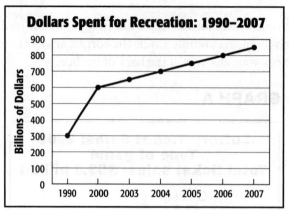

Dollars Spent for Recreation: 1990–2007

Source: U.S. Bureau of Economic Analysis, 2008

5. Graph B shows a steady increase in

   a. number of people who go to the movies
   b. percent of people who attend recreational events
   c. number of people who participate in recreational events each year
   d. money spent in admissions to amusement parks.
   e. money spent by consumers on all recreational activities

6. The percent of money spent for recreation between 1990 and 2000 _____.

   a. rose by 50%
   b. quadrupled
   c. doubled
   d. rose by 20%
   e. tripled

7. If the money spent on recreation continued at the same pace as it has from 2003 to 2007, one can expect the dollars spent on recreation in the year 2010 to be approximately _____.

   a. $950 billion
   b. $1 trillion
   c. $850 billion
   d. $900 billion
   e. none of the above

8. Which statement is supported by the contents of Graph B?

   a. More and more people are participating in outdoor recreational events.
   b. More households enjoy movies and plays today compared to those in the 1990's.
   c. About half of Americans will spend more money on recreation while on vacation.
   d. The amount of money spent on recreation will likely continue to increase over the years.
   e. People have more expendable income for leisure and recreation than for non-recreational activities.

GRAPH C

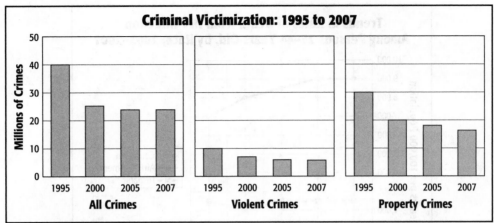

Criminal Victimization: 1995 to 2007

Millions of Crimes

All Crimes — 1995, 2000, 2005, 2007

Violent Crimes — 1995, 2000, 2005, 2007

Property Crimes — 1995, 2000, 2005, 2007

Source: U.S. Census Bureau

9. Graph C shows the number of _____ and _____ crimes for a 12-year period.

   a. private; total
   b. violent; property
   c. violent; rape
   d. burglary; violent
   e. gun; property

10. Total crimes during this 12-year period have dropped by approximately

   a. 30–35 million
   b. 20–25 million
   c. 17–18 million
   d. 10–12 million
   e. 8–10 million

11. From 1995 to 2007, there was a _____ percent decrease in violent crimes.

   a. 10
   b. 75
   c. 25
   d. 45
   e. 33

12. All of the following statements can be concluded from Graph C EXCEPT:

   a. Crime in all categories show a consistent decrease.
   b. There are more crimes committed against property than violent crimes.
   c. Of the years shown, 2005 and 2007 had the least total crimes.
   d. Crime has decreased by less than 10% during this 12-year period.
   e. Crimes have been decreasing over the years.

GRAPH D

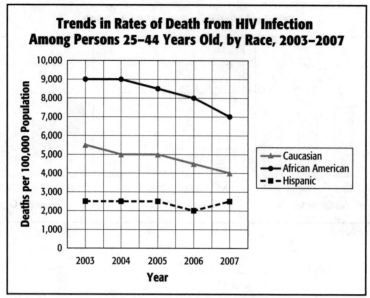

Trends in Rates of Death from HIV Infection Among Persons 25–44 Years Old, by Race, 2003–2007

Caucasian
African American
Hispanic

Source: Center for Disease Control and Prevention, 2009

13. Graph D shows the number of deaths per _____ population of individuals infected with HIV (AIDS).

    a. 25–44
    b. 1 million
    c. 100,000
    d. 200
    e. 20 million

14. The trend in HIV death rates for African Americans has consistently _____ since 2004.

    a. increased
    b. decreased
    c. doubled
    d. stayed the same
    e. tripled

15. In what years have Hispanics shown an increase in HIV-related deaths.

    a. 1997–2003   **c.** 2004–2005   **e.** 2006–2007
    b. 2003–2004   **d.** 2005–2006

16. Which statement is true according to Graph D?

    a. U.S. citizens show more HIV-related deaths than other groups.
    b. On the average, for every 180 deaths that occur in 100,000 African Americans, 10 to 15 deaths occur in 100,000 Caucasians.
    c. The trend in HIV-related deaths has been moderate decreases, with a few exceptions.
    d. More HIV-related deaths occur in the U.S. than in other countries.
    e. No HIV-related deaths have occurred among persons over 44.

SCHEDULE A

Schedule of Classes—Winter Quarter
Sociology (SOC)

Course Title	Course #	Meeting Time ‡	Location	Instructor
Internship/Practicum	410	TBA*	TBA*	Staff
Sociological Theory	413	TR 1100–1200	MLM 234	Conroy
Conducting Social Research	416	TR 1230–1350	MCC 201	Kenny
Rural-Urban Sociology	475	T 1800–2050	STAG 109	Kramer
Law and Society	491	TR 1230–1350	MLM 123	Sellers
Social Projects	506	TBA*	TBA*	Staff
Leisure and Culture	513	TR 1100–1220	MLM 234	Pershing
Conducting Social Research	516	TR 1230–1350	MCC 201	Plankford
Juvenile Delinquency	540	TR 1400–1520	STAG 111	Conway
Sociology of Religion	552	MWF 1300–1350	FAIR 305	Langley
Leisure and Culture	554	M 1900–2150	STAG 109	Mitchell
Gender Issues	575	T 1800–2050	STAG 109	Crailer

*TBA = to be announced
‡using a 24-hr clock, 1:30 P.M. would be 1330; M = Monday, T = Tuesday, W = Wednesday,
R = Thursday, F = Friday

17. Schedule A shows the range of courses that can be studied within what subject area?

   a. Spanish　　　**b.** Forestry　　　**c.** Law　　　**d.** Sociology　　　**e.** Religion

18. To register for the course called Juvenile Delinquency, one must register for what course number?

   a. Conway　　　**b.** 540　　　**c.** STAG 111　　　**d.** 410　　　**e.** 1400–1520

19. According to Schedule A, _____ and _____ are two courses whose meeting times are to be announced.

   a. SOC 410 and SOC 516
   b. SOC 506 and SOC 552
   c. SOC 410 and SOC 575
   d. SOC 410 and SOC 506
   e. SOC 416 and SOC 516

20. According to Schedule A, which course meets in the evening hours on Monday?

   a. SOC 554
   b. SOC 513
   c. SOC 475
   d. SOC 575
   e. SOC 491

CHART A

Affordable Current Monthly Premiums						
Issue Age	**$10,000**		**$20,000**		**$50,000**	
	Female	**Male**	**Female**	**Male**	**Female**	**Male**
45–49	$8.93	$11.42	$13.87	$18.83	$28.67	$41.08
50–54	10.46	13.41	16.92	22.82	36.29	51.04
55–59	14.38	18.12	24.77	32.23	55.92	74.58
60–64	20.44	24.85	36.88	45.70	86.21	108.25
65–69	27.38	31.99	50.75	59.98	120.88	143.96
70–74	39.83	44.65	75.65	85.30	183.13	207.25

Source: AARP Life Insurance Program from New York Life, 2010

21. Chart A shows the average monthly premiums for _____
_____.

    a. home flood insurance
    b. car, truck insurance
    c. term life insurance
    d. personal health insurance
    e. car, driver insurance

22. It is more cost effective to buy term life insurance when you are

    a. older
    b. retired
    c. female
    d. male
    e. younger

23. If a person wishes to buy a $20,000 term life insurance policy, what would is cost for a woman aged 57 and a man aged 62, respectively?

    a. $24.77 and $45.70
    b. $24.77 and $32.23
    c. $45.70 and $55.92
    d. $24.77 and $36.88
    e. $24.77 and $86.21

MAP A

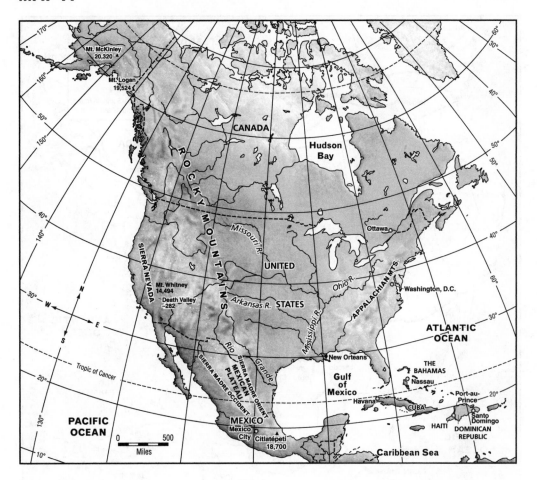

24. The capital city of Haiti is

    a. Santo Domingo
    b. Port-au-Prince
    c. Havana
    d. Nassau
    e. Bahamas

25. The Rocky Mountains run from _____ through _____.

    a. Canada; the U.S.
    b. Mexico; the U.S.
    c. Alaska; the U.S.
    d. Mexico; Canada
    e. Canada; Panama

26. All the statements below are true EXCEPT:

    a. Mexico City is 1,500–2,000 miles from Washington, D.C.
    b. All countries in North America are mountainous.
    c. A river acts as a boundary between Mexico and the U.S.
    d. Mt. Whitney is the highest mountain in North America.

MAP B

Creating a NAFTA Railroad
(North Atlantic Free Trade Agreement)

Legend:
— Canadian National
  Illinois Central
— Kansas City Southern routes,
  trackage rights and affiliates

27. Map B shows the NAFTA Railroad extending to cities as far north as _____, Canada, and as far south as _____, Mexico.

   a. Toronto; Mexico City
   b. Prince Rupert; Veracruz
   c. St. Louis; Detroit
   d. Vancouver; Halifax

28. All the statements below are true EXCEPT:

   a. The proposed NAFTA Railroad will be operated by the Canadian National Illinois Central and Kansas City Southern.
   b. The proposed NAFTA Railroad travels through three countries.
   c. The proposed NAFTA Railroad concentrates travel in east and west Canada, in central U.S., and in east Mexico.
   d. The proposed NAFTA Railroad travels through the largest cities of Canada, the U.S., and Mexico.

MAP C

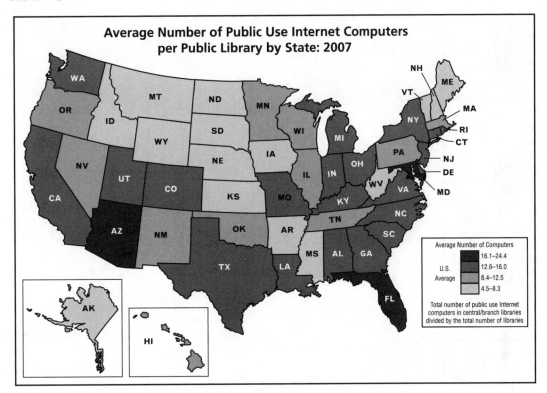

Average Number of Public Use Internet Computers per Public Library by State: 2007

Average Number of Computers
- 16.1–24.4
- 12.6–16.0
- U.S. Average 8.4–12.5
- 4.5–8.3

Total number of public use Internet computers in central/branch libraries divided by the total number of libraries

29. Map C shows the average number of public use _____ per _____ by state.

    a. internet browsers; public schools
    b. computer laptops; public agencies
    c. internet computers; public library

30. As of 2007, which states have the highest number of internet computers per public library?

    a. Alaska, Arizona, Florida
    b. Arizona, Maryland, Florida
    c. Maine, Arizona, Florida
    d. Georgia, Texas, Alabama
    e. Maryland, Florida, New Jersey

31. It can be inferred from Map C that there are more public internet computers in the public libraries of _____ than in the state of _____.

    a. Missouri; Arizona
    b. Louisiana; Virginia
    c. Alaska; Mississippi
    d. Nebraska; Louisiana
    e. Louisiana; Alaska

32. Which graph best describes the number of states on Map C?

a.
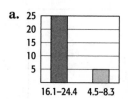

b.
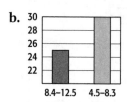

c.
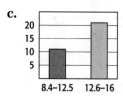

d.
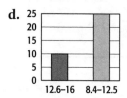

e.
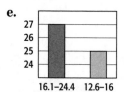

ANSWER KEY

Pages 1–9, Pretest

1. manufacturers; 1 million
2. False
3. a
4. c
5. six; death
6. False
7. b
8. d
9. population growth; 2030
10. False
11. c
12. b
13. high school students; 2007
14. False
15. b
16. e
17. average unemployment
18. True
19. b
20. c
21. Chicago, IL; New Orleans, LA
22. True
23. d
24. d
25. Asian Americans
26. True
27. d
28. d
29. Allegheny, Monongahela, Ohio
30. False
31. d
32. c
33. Mexico; Canada
34. False
35. c
36. a

Pages 12–15, Graph Skills Inventory

1. Southwest
2. 13,500,000 or 13.5 million
3. True
4. False
5. c
6. b
7. widowed; single
8. 8.7
9. True
10. True
11. c
12. b
13. states
14. $40,000
15. False
16. False
17. e
18. b
19. 600,000
20. True
21. b

Pages 22–23

1. a. Average Hourly Earnings in Different Job Categories
   b. Professional and Business Services; Wholesale and Retail Trade; Construction; Manufacturing; Information; Finance, Insurance, & Real Estate
2. $3.00
3. Construction
4. False
5. True
6. False
7. a
8. d
9. b
10. d

Pages 24–25

1. U.S. Department of Labor
2. 2004; 2009
3. percent
4. False
5. True
6. False
7. a
8. b
9. d
10. c

Pages 26–27

1. projected
2. 10 million
3. women; men
4. False
5. False
6. True
7. a
8. c
9. b
10. b

Pages 28–29

1. 2016
2. 450,000
3. True
4. False
5. True
6. d
7. a
8. d
9. c
10. a

Pages 30–31

1. membership
2. painters and allied professionals
3. True
4. False
5. True
6. c
7. a
8. d
9. c
10. a

Pages 34–35

1. a. The Federal Government Dollar: Where It Goes

   b. National Defense; Social Security; Net Interest; Medicare; Medicaid; Non-defense Discretionary; Reserves; Other
2. Social Security
3. National Defense
4. False
5. False
6. False
7. c
8. e
9. c
10. c

Pages 36–37

1. motor vehicles
2. pedestrian
3. 5
4. False
5. False
6. True
7. d
8. c
9. b
10. d

Pages 38–39

1. estimated
2. 322; 292
3. Housing; Health; Defense; Social Services
4. True
5. False
6. True
7. a
8. c
9. d
10. b

Pages 40–41

1. cancer
2. prostate
3. False
4. True
5. False
6. c
7. e
8. a
9. d
10. b

Pages 42–43

1. civilian labor force
2. b
3. True
4. False
5. True
6. b
7. c
8. d
9. b
10. e

Pages 46–47

1. a. Average Weight for Women 5' 7" Tall
   b. Average Weight (in lb)
   c. Age Groups
2. 20; 24
3. 158
4. True
5. False
6. False
7. c
8. c
9. d
10. c

Pages 48–49

1. average weight
2. 192 lb
3. 20; 69
4. False
5. True
6. True
7. e
8. a
9. e
10. b

Pages 50–51

1. men; women
2. 20, 24; 70, 79
3. 160 lb
4. True
5. False
6. True
7. d
8. a
9. d
10. d

Pages 52–53

1. third party; private consumer
2. one trillion, 450 billion
3. False
4. True
5. False
6. d
7. b
8. c
9. c
10. c

Pages 54–55

1. fatally injured
2. 40
3. True
4. False
5. d
6. b
7. b
8. a
9. d

Pages 58–59

1. whole milk
2. dollars; gallon
3. 2001; 2010
4. False
5. True
6. True
7. a
8. c
9. d
10. e

Pages 60–61

1. premium unleaded
2. cost; gallon
3. 20
4. False
5. False
6. True
7. c
8. b
9. d
10. e

Pages 62–63

1. Caucasians; African Americans; Hispanics
2. 1990; 2006
3. Hispanics
4. False
5. True
6. False
7. a
8. e
9. b
10. c

Pages 64–65

1. 17
2. 5%
3. True
4. False
5. False
6. b
7. b
8. c
9. d
10. a

Pages 66–67

1. electricity; coal; gas
2. electricity; coal
3. False
4. False
5. True
6. b
7. a
8. a
9. c
10. e

Pages 68–75, Graph Review

1. full-time workers
2. life earnings/income
3. True
4. False
5. a
6. b
7. b
8. b
9. death; U.S.
10. 8.2%
11. False
12. True
13. a
14. c
15. d
16. e
17. United States; 3
18. 238 million
19. False
20. True
21. c
22. b
23. c
24. a
25. beef; pork
26. 29; 2008
27. False
28. False
29. c
30. a
31. c
32. e

Pages 77–79, Schedule and Chart Skills Inventory

1. Central; Mountain
2. 3
3. False
4. True
5. d
6. zone; weight
7. 12
8. True
9. False
10. d
11. d

Page 87

1. vegetable; milk; meat
2. calories or servings
3. food
4. False
5. True
6. True
7. b
8. c
9. b
10. b

Pages 88–89

1. Johnson Community College
2. Mon., Jan. 26; Sat., Jan. 31
3. 787–1984
4. True
5. False
6. False
7. c
8. c
9. c
10. b

Pages 90–91

1. Suisun-Fairfield; Martinez; Richmond
2. 15; 18
3. sleeping car service
4. True
5. False
6. True
7. d
8. d
9. b
10. b

Page 92–93

1. actual
2. True
3. True
4. False
5. c
6. c
7. c
8. e
9. c

Pages 94–95

1. $49,000; $50,000
2. single; married filing jointly; married filing separately; head of household
3. True
4. False
5. True
6. a
7. a
8. c
9. e

Pages 96–101, Schedule and Chart Review

1. mileage
2. mileage; cities
3. True
4. False
5. a
6. c
7. e
8. b
9. depart; arrive
10. Portland; Los Angeles; Seattle; San Francisco
11. True
12. False
13. c
14. d
15. b
16. e
17. car; 5
18. $10,000; $30,000
19. True
20. True
21. b
22. c
23. d
24. c

Pages 102–105, Map Skills Inventory

1. mountains
2. latitude
3. False
4. True
5. c
6. downtown
7. east
8. False
9. True
10. c
11. d
12. annual rainfall; inches
13. True
14. False
15. d
16. d

Page 109

1. northwest
2. southeast
3. northeast
4. southwest
5. south central
6. southeast

Page 111

1. longitude
2. latitude
3. equator
4. 30° N
5. south

Page 113

1. D-3; H-7; D-5
2. Germany
3. Baltic Sea
4. Hamburg
5. Portugal
6. True
7. False
8. False

Page 115

1. a. 850   b. 450
   c. 400   d. 300
2. 200
3. D-5; 900; southeast
4. A-1; 850; northwest

Pages 120–121

1. Arctic Circle
2. 500
3. U.S.; Canada
4. 75° W
5. True
6. True
7. False
8. c
9. c
10. e

Pages 122–123

1. Guatemala; Honduras
2. Guatemala; Belize
3. Cuba
4. east
5. True
6. True
7. b
8. b
9. c
10. e
11. c

Page 127

1. San Francisco
2. A-4
3. east
4. False
5. False
6. True
7. True
8. c
9. b
10. b

Page 129

1. A-4
2. Springfield
3. Lake Michigan
4. False
5. True
6. False
7. c
8. a
9. e
10. c

Pages 132–133

1. 2010; Hispanic
2. California; Arizona
3. 500; 999
4. U.S. Census Bureau
5. False
6. True
7. False
8. d
9. a
10. d

Pages 134–135

1. home heating; five
2. electricity
3. A
4. False
5. True
6. b
7. a
8. b
9. c
10. d

Pages 137–141, Map Review

1. 300
2. Mekong
3. False
4. True
5. b
6. d
7. a
8. c
9. 6
10. A-7
11. False
12. True
13. d
14. b
15. c
16. b
17. Home Ownership; 2010
18. 60% or less
19. True
20. False
21. c
22. e
23. b
24. b

Pages 142–150, Posttest A

1. company; volume; 1 (one) million
2. False
3. d
4. d
5. unemployed
6. False
7. b
8. c
9. 18
10. False
11. c
12. d
13. Northeast
14. False
15. a
16. e
17. 85A; North Central
18. False
19. e
20. d
21. Agriculture
22. False
23. a
24. c
25. Colorado; Arizona
26. True
27. a
28. b
29. G-3
30. True
31. b
32. c
33. residential electric energy rate
34. False
35. d
36. a

Page 154–155

1. True   3. e

2. False   4. c

5.

INCOME ($5,000) – EXPENSES		
Expenses	Adjustments (if any)	Corrected Budget
Food	$800	0	$800
Clothing	$500	–$250	$250
Insurance	$550	0	$550
Housing	$1,600	0	$1,600
Medical	$400	–$100	$300
Transportation	$500	0	$500
Recreation	$250	0	$250
Savings	$250	0	$250
Credit Cards	$350	0	$350
Education	$300	–$150	$150
Total Expenses	$5,500	–$500	$5,000

Pages 156–157

1. $187,100   4. a

2. Denver   5. Yes; no; no; no; yes

3. d

Pages 158–159

1. $30

2. $210; $180; $90; $260; $165

3. $99; $143; $99

4. d

Page 161

1. $109.13; $17.34

2.

Monthly Access Charges	Wireless Surcharges
Phone #1: ($102.98)	Phone #1: ($6.15)
Phone #2: ($14.99)	Phone #2: ($2.35)
Total: ($117.97)	Total: ($8.50)

3. a

Pages 162–163

Note: Your answers may differ slightly from these. Approximations can vary.

1.

	1997		2010
Beef	9.00	Beef	12.00
Milk	6.40	Milk	7.20
Tomatoes	3.20	Tomatoes	3.60
Total	18.60	Total	22.80

2. $4.20

3. Tomatoes and milk

4. Beef: $43.20; Tomatoes: $7.20; Milk: $28.80; Total: $92.20

Pages 164–165

1. $(0.35 \times 20,000) + (125 \times 12) = 7,000 + 1,500$
$= \$8,500$

2. $5,500

3. c

4. $2,000

5. a. a small sedan with 20,000 miles
   b. $0.05
   c. i. True
      ii. True

Pages 166–167

1. a. 6-7 days   b. 3-4 days

2. $10,000; $4,383

3. $2,144

4. a. False   b. True   c. True

Pages 168–169

1.

From	To	Distance in Miles
Seattle	San Francisco	808
San Francisco	Los Angeles	379
Los Angeles	Houston	1,538
	Total Miles:	2,725

2. From	To	Distance in miles	Time in hours (speed = 50 miles per hour)
Seattle	Minneapolis	1608	32
San Francisco	Denver	1256	25
Houston	Raleigh	1160	23
Boston	Miami	1514	30

3. 20.5 mpg; 57 mph

4. e

Page 171

1. Sacramento; California

2. True

3. b

4. a. Sacramento Executive Airport (SAC)
   b. ARCO Arena
   c. California State University, Sacramento

5. a

Page 172

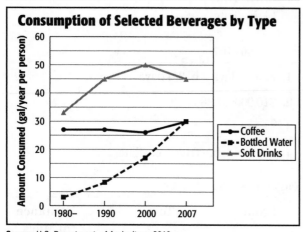

Consumption of Selected Beverages by Type

Source: U.S. Department. of Agriculture, 2010

Page 173

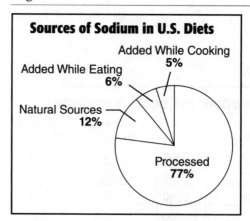

Sources of Sodium in U.S. Diets

Page 174

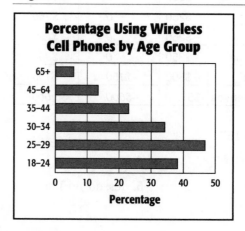

Percentage Using Wireless Cell Phones by Age Group

Page 175

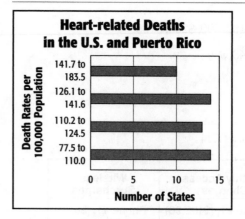

Heart-related Deaths in the U.S. and Puerto Rico

Page 176

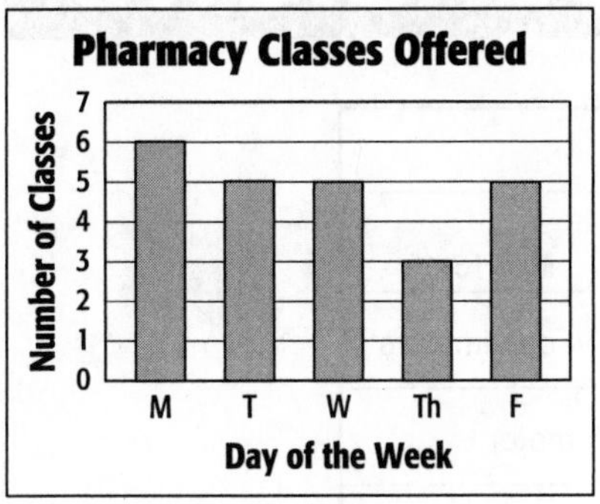
Pharmacy Classes Offered
Number of Classes
0
1
2
3
4
5
6
7
M
T
W
Th
F
Day of the Week

Pages 177–186, Posttest B

1. d	**9. b**	**17. d**	**25. a**
2. b	**10. c**	**18. b**	**26. d**
3. c	**11. d**	**19. d**	**27. b**
4. c	**12. d**	**20. a**	**28. d**
5. e	**13. c**	**21. c**	**29. c**
6. c	**14. b**	**22. e**	**30. b**
7. b	**15. e**	**23. a**	**31. e**
8. d	**16. c**	**24. b**	**32. c**

MEASUREMENT CONVERSIONS

LENGTH

Customary	Customary	Metric
1 inch	$\frac{1}{12}$ foot or $\frac{1}{36}$ yard	2.54 centimeters
1 foot	12 inches or $\frac{1}{3}$ yard	0.3 meter
1 yard	36 inches or 3 feet	0.91 meter
1 mile	5,280 feet or 1,760 yards	1.61 kilometers

LIQUID

Customary	Customary	Metric
1 ounce	$\frac{1}{16}$ pint	29.57 milliliters
1 cup	8 ounces or $\frac{1}{2}$ pint	0.24 liter
1 pint	16 ounces or 2 cups or $\frac{1}{2}$ quart	0.47 liter
1 quart	2 pints or 4 cups or $\frac{1}{4}$ gallon	0.95 liter
1 gallon	8 pints or 4 quarts	3.79 liters

WEIGHT

Customary	Customary	Metric
1 ounce	$\frac{1}{16}$ pound	28.35 grams
1 pound	16 ounces	453.6 grams
1 ton	2,000 pounds	907.18 kilograms

TEMPERATURE

°C = (°F − 32) ÷ 1.8

°F = (°C × 1.8) + 32

GLOSSARY

A

abbreviation One or more letters that are used to represent longer words. Directions on a map are usually abbreviated: N stands for north, S stands for south, E stands for east, and W stands for west.

annual Referring to anything that occurs each year

average A typical or middle value of a set of values. The arithmetic average (mean) is found by adding the set and then dividing the sum by the number of values in the set.

Stacey received three math scores: 83, 78, and 91. Stacey's average score is 84.

$83 + 78 + 91 = 252$

$252 \div 3 = 84$

axes The sides of a graph along which data values or labels are written

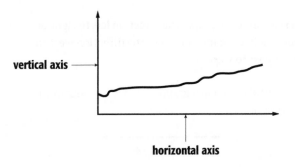

B

bar graph A graph that uses bars to display information. Data bars can be drawn vertically or horizontally.

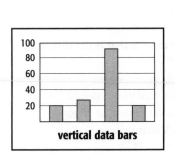 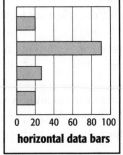

budget A financial plan that details expenses and sources of earning or revenue

C

chart A graph or drawing that contains data or other information

circle graph A graph that uses a divided circle to show data. Each part of the circle is called a *segment* or a *section*. The segments add up to a whole or to 100%. A circle graph is also called a *pie graph* or *pie chart*.

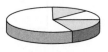

circle graph

column In a table, a vertical listing of numbers or words that is read from top to bottom

column of numbers	column of words
124	Newport
230	Jamestown
195	Cleveland
324	Detroit

continent One of the 7 major land masses of the earth. The continents are Antarctica, Africa, Asia, Australia, Europe, North America, and South America.

contour lines Lines on a topographical map that represent points of equal elevation

D

data A group of numbers or words that are related in some way

Number data: $2.50, $3.75, $6.40
Word data: beef, chicken, fish, pork

diagram Any chart or graph that contains numbers, words, or other information

direct distance The straight-line distance between two points as might be flown by an airplane

distance scale A map scale that relates ruler distance on a map to actual mileage

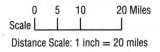

Distance Scale: 1 inch = 20 miles

double bar graph A bar graph containing two sets of bars to display and compare two sets of related data

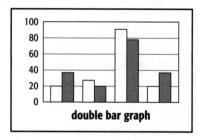

double bar graph

double line graph A line graph containing two lines to display and compare two sets of related data

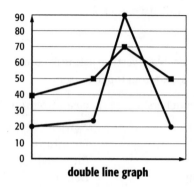

double line graph

directional map A map that is used to show the location of cities, highways, and points of interest. Directional maps are often called *road maps*.

E

east A map direction (**E**) that is opposite of west. When north is placed at the top of the direction symbol, east is placed at the right side.

Eastern Hemisphere The part of the earth that includes Asia, Africa, Australia, Europe, their islands, and all surrounding water

elevation key A symbol or number, often on a contour line, that tells land elevation

equator An imaginary line that circles the earth as a circumference, lying halfway between the North and South Poles. The equator is the line of 0° latitude.

extrapolate To make a reasonable guess of a data value that lies outside a given set of values

Four given data values: 5, 10, 15, 20
Extrapolated fifth value: 25

G

geographical map A map that shows the natural features of the earth. Natural features include lands, rivers, lakes, oceans, and other features that are not man-made.

globe A map in the shape of a sphere, showing the whole earth

globe

graph A pictorial display of information. A graph usually includes a graph title, data, horizontal and vertical axis labels, and the graph's source of information.

Greenwich, England A town in England through which the 0° longitude line passes. The longitude line numbered 180° is exactly halfway around the earth from Greenwich.

H

hemisphere On a map or globe, the word *hemisphere* refers to approximately half of the earth.

horizontal On a graph, the direction left to right or right to left. On a map or globe, the direction west to east or east to west

horizontal axis On a graph, the axis running from left to right

⟷

horizontal axis

I

index guide A set of map coordinates that is used to locate a point on a map. The coordinate of each point is usually given as a letter followed by a number—the letter giving the north/south location, the number giving the east/west location.

inference A guess at information that is not directly given

informational map A map that gives specific information about a particular region, or gives information comparing different regions

interpolate To estimate a data value that lies between two given values

K

key On a pictograph, an explanation of the meaning and value of a data symbol

L

latitude lines Lines drawn horizontally (west to east) across a globe. The latitude line number tells how far a point on the earth is north or south of the equator.

line graph A graph that displays data as points along a graphed line

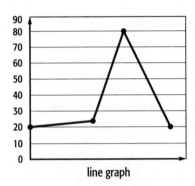

line graph

longitude lines Lines drawn vertically (north and south) on a globe. A longitude line number tells the east/west location of a point on the earth.

M

map A visual display that represents a city or a region of the earth

Mercator projection A flat map of the whole earth on which longitude lines are drawn as parallel lines. Exact locations are shown but sizes become increasingly distorted with distance from the equator.

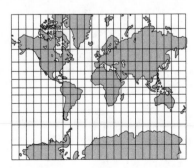

N

north A map direction (N) opposite of south. North is usually placed at the top of the direction symbol.

Northern Hemisphere The half of the earth that lies north of the equator

North Pole The northern most point on the earth. On a globe, the North Pole is at the top of the globe.

northeast Any direction that is both north and east of a point

northwest Any direction that is both north and west of a point

O

ocean The large body of salt water that covers most of the surface of the earth. The ocean is often geographically divided into five major oceans: Antarctic, Arctic, Atlantic, Indian, and Pacific. The Pacific Ocean borders the western shore of the United States; the Atlantic Ocean borders the eastern shore.

P

percent Part of 100. For example, 5 percent means 5 parts out of 100, 5 hundredths of the whole.

pictograph A graph that uses small pictures or symbols to represent data. Data lines may be displayed either horizontally or vertically.

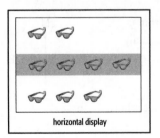
horizontal display

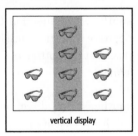
vertical display

pie graph *See circle graph.* (Pie graph is another name for circle graph.)

prediction Regarding a graph, a guess about the value of an unknown data point

R

road distance The actual road mileage distance between two points

row In a table, a horizontal list of numbers or words that is read from left to right

row of numbers: 26, 32, 65, 124
row of words: carrots, celery, lettuce, sprouts

S

schedule A list of facts and relations primarily dealing with times of events

City Center Bus Station Weekend Time Schedule			
Saturday	8:00 A.M.	10:30 A.M.	1:00 P.M.
Sunday	9:00 A.M.	11:30 A.M.	2:00 P.M.

sea level The average level of the ocean from which altitude measurements are compared. Land that is the same level as average level of the ocean is said to have a *zero elevation.*

segment On a circle graph, any pie-shaped division

← segment

source On a schedule, chart, graph, or map, a telling of where information was obtained

south A map direction (S) that is opposite of north. South is usually placed at the bottom of the direction symbol

southeast Any direction that is both south and east of a point

Southern Hemisphere The half of the earth that lies south of the equator

South Pole The southern most point on the earth. On a globe, the South Pole is at the bottom of the globe.

southwest Any direction that is both south and west of a point

T

table A display of data (words and numbers) organized in rows and columns

Name of School	Women Teachers	Men Teachers
Jefferson	14	6
Lincoln	12	9
Washington	17	4

row —→ Lincoln

column

title On a graph, a short description of the main information presented

V

vertical axis On a map, the axis running up and down

vertical axis

W

west A map direction (W) that is opposite of east. If north is placed at the top of the direction symbol, west is placed at the left side.

Western Hemisphere The part of the earth that includes North and South America, their islands, and the surrounding water

INDEX